Elements of Distributed Algorithms

T0185175

Springer
Berlin
Heidelberg
New York
Barcelona
Budapest
Hong Kong
London
Milan
Paris
Singapore
Tokyo

Wolfgang Reisig

Elements
of Distributed Algorithms

Modeling and Analysis
with Petri Nets

With 246 Figures

Springer

Professor Dr. Wolfgang Reisig

Humboldt-Universität zu Berlin
Institut für Informatik
Unter den Linden 6
D-10099 Berlin, Germany

E-mail: reisig@informatik.hu-berlin.de

Library of Congress Cataloging–in–Publication Data

Reisig, Wolfgang, 1950–
    Elements of distributed algorithms: modeling and analysis with
    Petri nets/Wolfgang Reisig.
        p. cm.
    Includes bibliographical references.

    1. Electronic data processing–Distributed processing.
    2. Computer algorithms. 3. Petri nets.   I. Title.
    QA76.9.D5R445   1998
    004'.36–dc21   98-22854
                                                                     CIP

ISBN 978-3-642-08303-7

© Springer-Verlag Berlin Heidelberg 2010
Printed in Germany

The use of general descriptive names, trademarks, etc. in this publication does not imply,
even in the absence of a specific statement, that such names are exempt from the relevant
protective laws and regulations and therefore free for general use.

Cover Design: Künkel + Lopka, Werbeagentur, Heidelberg

Preface

The application and organization of computing systems is tending towards distributed computing. Processor clusters, local and wide area networks, and the forthcoming information highway are evolving new kinds of problems (the simplest and most basic ones include, e.g., distributed organization of mutual exclusion, or distributed detection of distributed termination). A new kind of algorithm, called *distributed algorithms*, has emerged during the last decade, aimed at efficiently solving those problems and, more generally, making distributed computing systems applicable to real-world problems.

A variety of distributed algorithms are presented and proven correct in this book. A (Petri net based) technique to model and to analyze distributed algorithms is coincidently presented. This technique focuses on local states, independent actions, and synchronization of distributed threads of control.

This book's scope is modest, as it sticks to a choice of small and medium size distributed algorithms. Compositionality, stepwise refinement, interface specification, abstraction, etc., are not covered. Nevertheless, this book's claims are ambitious: Just as PASCAL-like programming structures and Hoare-style proof techniques appear optimal for a wide class of sequential algorithms, this book's formalism is suggested to be optimal for a wide class of distributed algorithms.

Particular preliminary knowledge is not assumed in this text, besides basics in formal concepts and a general intuitive understanding of computer science.

The text provides a basis for university courses and can help the practitioner to design distributed algorithms. The hurried reader may just study the pictures.

Acknowledgments

This book is the yield of a decade of research into formal methods for modeling and analysis of Distributed Algorithms. I conducted this research together with many colleagues, staff, and students, both at the Technical University of Munich and, since 1993, the Humboldt-Universität zu Berlin. It has been supported by the Deutsche Forschungsgemeinschaft, in the framework of the Sonderforschungsbereich 342 as well as in projects on Distributed algorithms design and compositional verification. The European Union supported this research via the ESPRIT II projects CALIBAN and DEMON.

The material matured and the presentation profited from successive courses given at the Technical University of Munich and Humboldt-Universität zu Berlin. My enjoyable summer 1997 visit to the International Computer Science Institute at Berkeley, California, decisively boosted completion of this book.

With pleasure I acknowledge a number of colleagues who directly contributed to this book: Rolf Walter suggested the concept of rounds and the corresponding theorems of Sect. 53. He also designed the crosstalk algorithms, crosstalk based mutex, and distributed rearrangement (Sects. 12, 13.6, and 25), and he contributed to token passing mutex, global mutex, and to the proofs of the Echo algorithm and phase synchronization (Sects. 13.5, 34.1, 77, and 81). Ekkart Kindler designed the fairness rules (Sects. 48 and 66). He constructed the spanning tree algorithm (Sect. 32.3) and contributed to token passing mutex, global mutex, and rearrangement (Sects. 13.5, 34.1, and 25). Jörg Desel contributed two conventions: how to syntactically represent network algorithms (Sect. 31) and how to handle equations, inequalities and logical expressions (Sect. 39). He also contributed to the global mutex algorithm (Sect. 34.1). Hagen Völzer constructed the phase synchronization algorithm together with its verification (Sects. 36 and 81) and contributed to the proof of the Echo algorithm (Sect. 77). Jörn Freiheit corrected an earlier version of local mutual exclusion on networks (Sect. 34.3) and contributed decisive arguments to its verification (Sect. 79). Tobias Vesper provided the central proof arguments for the distributed self-stabilization algorithm (Sect. 82). Wilfried Brauer constructed the asynchronous stack (Sect. 11). Eike Best referred me to [Dij78], the source of this book's consensus algorithms (Sect. 35).

In addition, the above-mentioned contributed numerous comments and corrections, as likewise did Juliane Dehnert, Adriane Foremniak, Dominik Gomm, Keijo Heljanko, Bodo Hohberg, Irina Lomazowa, Sibylle Peuker, Thomas Ritz, Karsten Schmidt, Einar Smith, and Michael Weber. Many thanks to all of them, and to the many further casual members of the "Kaffeerunde", for inspiring discussions and criticism.

The burden of never-ending typesetting, changes of text, and corrections was shouldered mainly by Thomas Ritz. I am very grateful for his patience and constructive cooperation. The pictures were drawn by Jörn Freiheit and Abdourahaman in quite a painstaking manner. Juliane Dehnert compiled the references. I am indebted to all of them.

I also owe much to Birgit Heene; her effective management of daily secretarial duties saved me a lot of time that I was able to devote to this book.

It is a pleasure to acknowledge the constructive, efficient, and friendly support of Hans Wössner, Ingeborg Mayer, and J. Andrew Ross of Springer-Verlag. Their professional management and careful editing of the manuscript have again been perfect, just as I frequently experienced during the last 15 years.

A Note to the Expert in Petri Nets

This is not a book on Petri nets. It is a book however that heavily *employs* Petri nets, i.e., a slightly revised form of elementary net systems (en-systems), and "algebraic" high level nets. Revisions are motivated by the realm of distributed algorithms: The role of loops is reconsidered, and aspects of progress and fairness are added. Altogether, those revisions provide the expressive power necessary to model elementary distributed algorithms adequately, retaining intuitive clarity and formal simplicity.

A version of high-level nets is used to model data dependencies and algorithms for (classes of) networks. Technically, such nets are based on concepts from algebraic specification, but they may be understood independently of that field.

Multiple instances of a token in the same place will never occur, because the vast majority of distributed algorithms by their nature do without multiple instances of tokens. Local states of nets are hence propositions and predicates, thus providing the elements for logic-based analysis techniques.

Two kinds of properties will be studied, both of them instances of [AS85]'s *safety* and *liveness* properties. In this text they will be called *state* and *progress* properties in order to avoid confusion with conventional *safety* and *liveness* of Petri nets. Again, those notions are motivated by practical needs. Place invariants and initialized traps turn out to be useful in analyzing state properties. Progress properties have rarely been considered in the literature on Petri nets. New concepts support their analysis, borrowed from temporal logic.

Berlin, April 1998   Wolfgang Reisig

Contents

Introduction 1

Part A. Elementary System Models   3

I.   **Elementary Concepts** 5
    1   A First Look at Distributed Algorithms 5
    2   Basic Definitions: Nets 14
    3   Dynamics 17
    4   Interleaved Runs 21
    5   Concurrent Runs 22
    6   Progress 27
    7   Fairness 30
    8   Elementary System Nets 31

II.   **Case Studies** 35
    9   Sequential and Parallel Buffers 35
    10   The Dining Philosophers 38
    11   An Asynchronous Stack 44
    12   Crosstalk Algorithms 45
    13   Mutual Exclusion.................................... 50
    14   Distributed Testing of Message Lines.................. 60

Part B. Advanced System Models   63

III. **Advanced Concepts** 65
    15   Introductory Examples 65
    16   The Concept of System Nets 73
    17   Interleaved and Concurrent Runs 75
    18   Structures and Terms 78
    19   A Term Representation of System Nets................ 80
    20   Set-Valued Terms 83
    21   Transition Guards and System Schemata 88

IV.   Case Studies 91
   22   High-Level Extensions of Elementary Net Models 91
   23   Distributed Constraint Programming 96
   24   Exclusive Writing and Concurrent Reading 100
   25   Distributed Rearrangement 102
   26   Self Stabilizing Mutual Exclusion 105

V.   Case Studies Continued: Acknowledged Messages 107
   27   The Alternating Bit Protocol 107
   28   The Balanced Sliding Window Protocol 112
   29   Acknowledged Messages to Neighbors in Networks 116
   30   Distributed Master/Slave Agreement 119

VI.   Case Studies Continued: Network Algorithms 123
   31   Principles of Network Algorithms 123
   32   Leader Election and Spanning Trees 125
   33   The Echo Algorithm 127
   34   Mutual Exclusion in Networks 130
   35   Consensus in Networks 134
   36   Phase Synchronization on Undirected Trees 137
   37   Distributed Self Stabilization 140

Part C. Analysis of Elementary System Models   143

VII.   State Properties of Elementary System Nets 145
   38   Propositional State Properties 145
   39   Net Equations and Net Inequalities 147
   40   Place Invariants of es-nets 150
   41   Some Small Case Studies 153
   42   Traps 156
   43   Case Study: Mutex 159

VIII. Interleaved Progress of Elementary System Nets 165
   44   Progress on Interleaved Runs 165
   45   The Interleaved Pick-up Rule 167
   46   Proof Graphs for Interleaved Progress 170
   47   Standard Proof Graphs 172
   48   How to Pick Up Fairness 176
   49   Case Study: Evolution of Mutual Exclusion Algorithms.... 178

IX.   Concurrent Progress of Elementary System Nets 187
   50   Progress on Concurrent Runs 187
   51   The Concurrent Pick-up Rule 188
   52   Proof Graphs for Concurrent Progress 190

53 Ground Formulas and Rounds............................ 191
54 Rounds of Sequential and Parallel Buffer Algorithms 195
55 Rounds and Ground Formulas of Various Algorithms....... 197
56 Ground Formulas of Mutex Algorithms.................. 200

Part D. Analysis of Advanced System Models   **205**

X.   State Properties of System Nets 207
57 First-Order State Properties 208
58 Multisets and Linear Functions......................... 209
59 Place Weights, System Equations, and System Inequalities . 210
60 Place Invariants of System Nets 214
61 Traps of System Nets 219
62 State Properties of Variants of the Philosopher System 222

XI.   Interleaved Progress of System Nets 227
63 Progress on Interleaved Runs 227
64 Interleaved Pick-up and Proof Graphs for System Nets 228
65 Case Study: Producer/Consumer Systems 230
66 How to Pick up Fairness................................ 231

XII. Concurrent Progress of System Nets..................... 233
67 Progress of Concurrent Runs............................ 233
68 The Concurrent Pick-up Rule 234
69 Pick-up Patterns and Proof Graphs 235
70 Ground Formulas and Rounds........................... 238

XIII. Formal Analysis of Case Studies 241
71 The Asynchronous Stack 241
72 Exclusive Writing and Concurrent Reading 244
73 Distributed Rearrangement 249
74 Self-Stabilizing Mutual Exclusion 253
75 Master/Slave Agreement 255
76 Leader Election 258
77 The Echo Algorithm 260
78 Global Mutual Exclusion on Undirected Trees............ 266
79 Local Mutual Exclusion 269
80 Consensus in Networks 279
81 Phase Synchronization on Undirected Trees 285
82 Distributed Self-Stabilization........................... 291

References 299

Introduction

An algorithm is said to be *distributed* if it operates on a physically or logically distributed computing architecture. Typically, such architectures lack global control. This requires particular means to model and to verify distributed algorithms.

This book is based on the assumption that distributed algorithms are important and that their present-day treatment can be improved upon. Two central problems are tackled: how to adequately *describe* a distributed algorithm and how to *prove* its decisive properties.

The algorithmic idea of most distributed algorithms centers around messages, synchronizing shared use of scarce resources, and causal dependencies of actions for some particular, usually not fully specified, computing architecture.

As an example, the *echo algorithm* is a schema for acknowledged information dissemination, to run on any connected network of computing agents. From a purely computational point of view, this algorithm just stipulates that each agent send the same message to all its neighbors. The *algorithmic idea*, however, is encoded in causal relations *between* those messages. Any adequate description of this idea should employ formal primitives to represent sending and receiving of messages; whereas, e.g., the administration of messages already received in an array is an implementation detail, irrelevant for the echo algorithm.

A distributed algorithm is adequately described if the operational primitives employed focus on the essential idea of the algorithm. Experience reveals that *local states* and *atomic actions* are crucial in this context: Occurrence of an atomic action affects a subset of local states. More involved examples furthermore require values to be moved along the system and require a more abstract description of computing systems (such as "any connected network of agents" in the echo algorithm). Technically, an adjusted version of Petri nets offers primitives of this kind.

The decisive properties of each distributed algorithm include aspects of *safety* and *liveness*, intuitively characterized as "nothing bad will ever happen" and "eventually something good will happen", respectively. As an example, the core properties of the echo algorithm mentioned above are, firstly, that the agent starting to disseminate some message will terminate only after all other agents have been informed and, secondly, that this agent *will* eventually terminate. Technically, temporal logic provides adequate means to represent and to prove these kinds of property.

Hence this book will employ an adjusted version of Petri nets to *represent* distributed algorithms, and an adjusted version of temporal logic to *verify* them. It combines selected concepts that reveal transparency and simplicity of both representation and analysis of distributed algorithms. These include

- suitable means to represent the essentials of distributed algorithms (such as local states, atomicity of actions, and synchronization), and to avoid unnecessary and superfluous concepts such as variables and assignment statements;
- a maximally tight combination of modeling and analysis techniques, where local states *are* propositional or first-order expressions, and actions *are* most elementary predicate transformers;
- well-established Petri net analysis techniques (*place invariants* and *initialized traps*, in particular), immediately yielding logical representations of safety properties (in the sequel called *state properties*);
- suitable means based on temporal logic to represent and prove liveness properties, by "picking up" elementary such properties from the static presentation of algorithms, and by combining them in *proof graphs* (in the sequel called *progress properties*);
- new notions of *progress* and *fairness* that differ slightly from conventional notions of *weak* and *strong fairness*, and yield amazingly simple proof rules.

We do not pursue the most expressive means available in an attempt to cover virtually *all* interesting aspects and properties of distributed algorithms. Instead, we restrict our attention to technically quite simple means, yet still covering an overwhelming majority of the problems that arise during the construction and analysis of distributed algorithms.

Special features of this text, besides its general aim of ultimate transparency and simplicity, include

- the notion of *concurrent runs*, as a basis for the notions of *ground states* and *round based* algorithms;
- a slightly revised notion of fairness;
- particularly strong techniques for picking up safety and liveness properties from the static representation of distributed algorithms.

Part A
Elementary System Models

This part consists of two chapters. The first one introduces the elementary ingredients for modeling distributed algorithms: local states, atomic actions, the notion of single runs, and the assumption of progress (among others). Altogether, those elementary concepts are amazingly simple. Nevertheless, they provide adequate means to model a large class of distributed algorithms.

The second chapter demonstrates this by various well-known distributed algorithms formulated in this setting. It turns out that crisp models frequently do better without variables, assignment statements, global fairness assumptions, etc.

I. Elementary Concepts

Different representations of an intuitively simple system are presented and compared in Sect. 1. One of them is a *net*, i.e., an instance of the formalism introduced in the rest of Chap. I. Section 2 is slightly technical as it introduces nets, the basic notion of this book. Most of Sect. 2 – anyway small – may be skipped upon first reading. The elementary formalism for modeling distributed algorithms is introduced in Sect. 3. It essentially comprises the notions of *local states*, *local actions*, and their interplay. Two respective conceptions of *single runs* of concurrent systems are considered in Sects. 4 and 5: a conventional one, employing global states and sequentially observed events, and a causality-based notion, emphasizing causal order and locality of state occurrences and events. Section 6 considers the – intuitively obvious – assumption of *progress*, i.e., the assumption that an enabled action will either occur or be disabled by some occurrence of competing action. The fundamentals of priority and fairness follow in Sect. 7. Section 8 concludes the chapter with remarks on the suggested techniques.

1   A First Look at Distributed Algorithms

This section provides a first look at this book's contents. We start by outlining the scope of concern, i.e., representations of distributed algorithms. Then different such representations are presented for a fairly simple, albeit illustrative example. Among them is a Petri net. Petri nets will be employed throughout this book.

1.1   Scope of concern

Distributed algorithms help to organize a large class of technical as well as non-technical systems. A system may exist physically or be implemented organizationally, or it may be a planned, hypothetical reality. Examples of systems include any kind of workflow (e.g., an office for issuing passports) and technical systems (e.g., production lines in a factory), and of course every kind of computing device.

A system is assumed to exhibit some quantum of *dynamic change*. Dynamic change is often described as a continuous function over time. However it is, frequently more appropriate to identify *discrete* change. For example, on the level of register-transfer in a computer, moving a value into a register is achieved by continuously changing voltages. A continuous model may describe physical details. The intended effect, however, i.e., what the hardware is built for and intended to provide, is nevertheless discrete. This kind of discrete effect will be modeled as an *action*.

Typical such actions of computer systems can be identified in logical switches, machine instructions, compiling and programming issues, database transactions, network organization, etc.

Examples of actions in organizational systems, e.g., in a passport office, include filling in an application form, delivering it to a clerk, paying a fee at a cash desk, receiving a receipt, etc.

Actions also describe a lot of relevant behavior in technical systems. Typical actions in chemical processes include heating some liquid, or pouring it into bottles. Of interest in this context are usually measuring instruments signaling, e.g., "liquid is hot enough" or "bottle is full".

Actions arise in the process of *modeling* behavior of systems, by help of algorithms. Many different *formalisms* for describing such algorithms have been suggested. This book employs the formalism of Petri nets.

1.2   Example: A producer-consumer system

In the rest of this chapter, a fairly simple, albeit illustrative distributed algorithm will be modeled, in the sequel denoted as a *producer-consumer system*. This algorithm may originate from quite different areas, including databases, communication protocols, operating systems, or computer architecture (but also from areas outside computer science): Distinguished items are produced, delivered to a buffer, later removed from the buffer, and finally consumed. The buffer is assumed to have capacity for one item. In a concrete instance, "to deliver" may stand for "to send" or "to deposit". Likewise, "to remove" may stand for "to receive" or "to accept". The items involved may be any goods, data carries, signals, news, or similar items. We are not interested here in any particular one of those concrete instances, but in their common properties.

Four models of this algorithm will be studied, using four different formalisms. We start with a programming notation, continue with state and transition based formalisms, and finally construct a Petri net.

1.3   A programming notation for the producer-consumer system

One may be tempted to represent the producer/consumer system in terms of programming notations. As an example, assume a variable *buffer*, ranging over the two-element domain {empty, filled}, and the following two programs:

P1: **do** forever
   produce;
   if buffer = empty **then**
     buffer := filled
   end

P2: **do** forever
   if buffer = filled **then**
     buffer := empty;   (1)
   consume
   end.

Furthermore, let $P = (P_1 \parallel P_2)$ be a program, representing the parallel execution of P_1 and P_2.

In order to clarify the meaning of P, one may assume a compiler, a run time system, etc., allowing one to execute P on a given computing device. The runs of P then correspond to the behavior of the producer/consumer system: The system is *simulated* by P.

1.4 A state-based representation of the producer-consumer system

As a second approach to represent the system described above, we may assume three subsystems: the *producer*, the *buffer*, and the *consumer*, with the following states and actions:

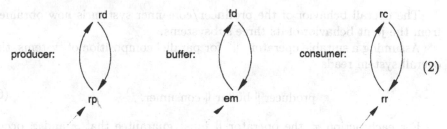

Each subsystem consists of two local states which are visited alternately: the producer's states are "ready to produce" (*rp*) and "ready to deliver" (*rd*). Likewise, the buffer has the states "empty" (*em*) and "filled" (*fd*), and the consumer's states are "ready to receive" (*rr*) and "ready to consume" (*rc*). Arrows denote possible steps between the subsystems' states. Arrows without a source node indicate initial states.

Diagram (2) does not describe the behavior of the system entirely and correctly: The steps $rd \to rp$ and $em \to fd$ occur coincidently in the system. This applies also for $rr \to rc$ and $fd \to em$. This information must be given in addition to (2).

1.5 An action-based representation of the producer-consumer system

The above representation is *state-based*: (local) states are assumed, and transitions are described as functions over these states. Next we consider an example for an *action-based* approach. The three subsystems are now described by their actions: The producer alternates the actions "produce"

(p) and "deliver" (d). The infinite sequence pdpd... of actions describes its behavior. This behavior can be represented finitely by the equation

$$\text{producer} = \text{p.d.producer.} \tag{3}$$

This equation may be read as "the producer first performs p, then d, and then behaves like the producer". The infinite sequence pdpd... is the solution of the equation (2) in the set of all (finite and infinite) strings consisting of p and d.

The consumer likewise alternates the actions "remove" (r) and "consume" (c). Its infinite behavior rcrc... is given by the equation

$$\text{consumer} = \text{r.c.consumer.} \tag{4}$$

The buffer's actions are strongly synchronized with depositing and removing, and are therefore denoted by $\bar{d}$ and $\bar{r}$, respectively. Its behavior $\bar{d}\bar{r}\bar{d}\bar{r}$... is given by

$$\text{buffer} = \bar{\text{d}}.\bar{\text{r}}.\text{buffer.} \tag{5}$$

The overall behavior of the producer/consumer system is now obtained from the joint behavior of its three subsystems.

Assuming a suitable operator "||" for parallel composition of systems, the overall system reads

$$\text{producer} \parallel \text{buffer} \parallel \text{consumer.} \tag{6}$$

For each action x, the operator || must guarantee that x and $\bar{x}$ occur *coincidently*.

1.6 A net representation of the producer-consumer system

The two formalisms of Sects. 1.4 and 1.5 operate with quite fixed concepts: One is state-based, one is action-based, and both stick to pairwise synchronization of actions of sequential processes. With these kinds of formalism, a lot of distributed algorithms can be represented adequately. But we are after a *basic, neutral* formalism, treating states and actions on an equal footing, and avoiding the need to fix particular decompositions of algorithms already at the beginning.

Figure 1.1 represents the producer/consumer system as a *Petri net*. It employs *circles* and *boxes* to represent local states and actions, respectively. Black dots ("tokens") inside circles characterize the initial state of the system. Generally, tokens indicate local states that are "presently taken" or "reached". Each of the four involved actions may *occur* under certain circumstances, thus changing the actual distribution of tokens.

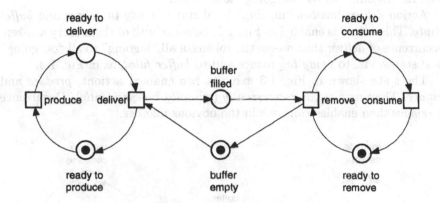

Figure 1.1. A net model of the producer/consumer system

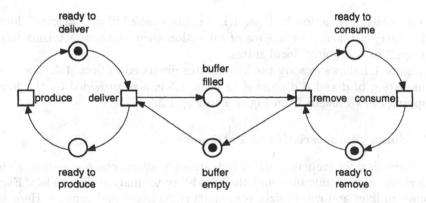

Figure 1.2. After occurrence of *produce*

Action *produce* is *enabled* whenever *ready to produce* carries a token. Occurrence of *produce* forwards the token to *ready to consume*, thus producing the global state of Fig. 1.2. Intuitively, occurrence of *produce* moves a token from its "ingoing" to its "outgoing" local state.

Action *deliver* has *two* "ingoing" local states, *ready to deliver* and *buffer empty*. This action is enabled in Fig. 1.2, because both of them carry a token. Occurrence of *deliver* then moves the tokens of all "ingoing" to all "outgoing" local states, viz. to *ready to produce* and to *buffer filled*, as in Fig. 1.3.

The state shown in Fig. 1.3 exhibits *two* enabled actions, *produce* and *remove*. They may occur *concurrently* (*causally independently*). Occurrence of *remove* then enables *consume* in the obvious manner.

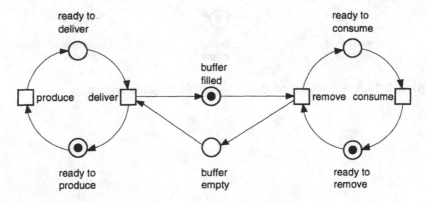

Figure 1.3. After occurrence of *deliver*.

Generally, any action in Figs. 1.1 – 1.3 is *enabled* iff all "ingoing" local states carry a token. Occurrence of an action then moves the tokens from "ingoing" to "outgoing" local states.

Figure 1.1 shows exactly the local states discussed in Sect. 1.4. The synchronization of d and $\bar{d}$ discussed in Sect. 1.5 is now provided by the joint swapping of two tokens from Fig. 1.2 to Fig. 1.3.

1.7   Some characteristics of Petri nets

Any formalism for modeling algorithms strongly affects comprehension of the underlying algorithmic idea and the capability to analyze the models. Each formalism fixes and emphasizes some particular issues and aspects. Here we discuss six such characteristics of the formalism considered in this book.

Petri nets (in their elementary setting) employ the most elementary concepts of states and transitions.   (7)

A *local state* is just a logical (propositional) atom (e.g. "ready to produce" in Fig. 1.1) with a truth value that may change upon an action's occurrence. An action consists (as in many formalisms) of an enabling condition and an effect on states upon its occurrence. Based on the local states introduced above, the logical conjunction of some local states (with their actual truth values) serve as an action's enabling condition. The effect of an action's occurrence consists in swapping the actual truth value of some local states. All occurrences of an action cause the same local effect.

Furthermore, actions *preserve information*: From an action and the result of its occurrence, the previous state can be recomputed. This strongly supports analysis techniques.

The propositional character of local states will be generalized to predicates in Part B, thus reflecting data values and schematic descriptions of systems.

Petri nets emphasize locality of causation and effect.   (8)

Each local state is assigned a fixed set of actions that swap (or read) its truth value. Likewise, each action is assigned a fixed set of local states that are involved in its occurrence. The graphical representation of nets (such as in Fig. 1.1) represents this vicinity explicitly (by arrows).

Petri nets explicitly represent fundamental issues of distributed systems, such as atomicity, synchronization, mutual independence   (9)
of actions, messages, and shared memory.

Conventional control flow, conventional variables, and assignment statements are no basic features of Petri nets. They can nevertheless be simulated, but proper Petri net representations of distributed algorithms do without those features.

Petri nets are neither state-based nor action-based. Both states
and actions have a particular status on their own.   (10)

This implies that Petri nets can be employed as a reference formalism for many kind of modeling formalism. In fact, the semantics of various concurrent languages has been formulated in terms of Petri nets.

Petri nets are unstructured by definition. Structure may be put
on them additionally.   (11)

This may be considered a disadvantage of Petri nets: Compositional proof techniques, algebraic representations, and any kind of inductive arguments can not be applied immediately. However, (11) is also advantageous. As an example, there exist standard techniques to gain the three components (producer, buffer, consumer) of the net in Fig. 1.1 (they are just *place invariants*). But a different decomposition may likewise help: The "upper line" from left to right, carrying produced items, and the "lower line" from right to left, carrying "empty" signals.

Petri net models are implementable and are neutral against specific implementation languages.   (12)

This of course is due to (10) and (11).

To sum up, Petri nets *abstract from specific concepts* such as state orientation, event orientation, pairwise synchronization, composability from sequential components, variables, values, assignment statements, hierarchical structuring, and similar concepts. What remain are *fundamentals* of models for distributed algorithms. This view is neither particularly "low" or particularly "high", it is just a particular level of abstraction. The remaining concepts are taken from logic: The conventional technique of assigning predicates to (global or local) states has in Petri nets been strengthened to taking propositions and predicates themselves as local states.

It will turn out that this provides an amazingly useful modeling formalism. It includes a kind of insight into distributed algorithms and properties that is hard to find or present by other means.

1.8   Relationship to other formalisms

The conception of actions, as introduced in Sect. 1.1 and discussed under characteristic (7), is employed in many other formalisms. Examples include guarded commands [Dij75], UNITY [CM88], and extended transition systems [MP92]. But emphasis on local states is typical for Petri nets. Other formalisms usually work with global states.

Use of logical concepts (propositions in Part A and predicates in Part B of this book) as local states is also a specific issue. Other formalisms use states to store values of variables and flow of control. They employ predicates on states whenever properties of models are to be analyzed. Employing predicates as state elements is hence a natural choice.

Reversibility of actions is motivated by fundamental considerations in, e.g., [FT82] and [Ben73]. They show relevance and universality of reversible switching elements.

Locality of causation and effect, as described in (8), is a fundamental issue [Gan80]. This contrasts with other formalisms such as *CCS* [Mil89] and *state charts* [Har87], where in order to fully conceive an action, one has to trace the occurrence of a distinguished symbol in the entire system model (i.e., in a term or a graphical representation).

Compared with Petri nets, other formalisms cope with fundamental issues of distributed algorithms less explicitly. As an example, assumptions on atomicity of parts of assignment statements are frequently not made explicit, though they are crucial for the semantics of parallel programs. Synchronization issues such as "wait for messages from all your neighbors, and then ... " are formulated implicitly, e.g., by counting the number of arriving messages.

Many other modeling formalisms are compositional and define system models inductively, e.g., CCS [Mil89] and state charts [Har87]. Such models suggest, but also generally fix, compositional and hierarchical proof structures, term representations, and any kind of inductive arguments. Unstructured formalisms such as Petri nets and transition systems are better at allowing for example oriented, goal-guided decomposition and structuring when it comes to correctness proofs.

This book's view can not be retained for large systems. Systematic refinement of specifications and compositional proof techniques are inevitable then. However, large algorithms require adequate techniques for *small* algorithms. The sequel is intended to provide such techniques, in particular providing simpler means and arguments for a wide class of distributed algorithms.

1.9   Relationship to other textbooks

One of the earliest textbooks is [CM88], providing a simple abstract operational model, a temporal logic based proof technique, and an impressive collection of case studies. In the latter two respects, the scope of this book almost coincides with ours. But we employ a fundamentally different operational model, which explicitly models concurrency (as well as nondeterminism) and is implementable in principle, as it refrains from global fairness assumptions. Fred Schneider's most recent book [Sch97] suggests conventional concurrent programs and adjusted, well-established temporal logic-based proof techniques. Concurrency is operationally treated as a special case of nondeterminism, and fairness assumptions affect global states (both contrasting with our basic assumptions).

Some issues treated in [BA90] and in our book coincide, including algorithms for (distributed) mutual exclusion and dining philosophers. [BA90] concentrates on programming concepts, specific programming languages, and implementation strategies, whereas we concentrate on an abstract implementable operational model and on verification.

[RH90] discusses a lot of synchronizing algorithms, some of which we pick up, too. [RH90] represents algorithms in semi-formal pseudo code, where we use a formal operational model. We give the notion of a "wave", suggested in [RH90], a formal basis, and exploit it in proof techniques.

In the line and style of [BA90], [Ray88] generalizes that approach to other algorithms, and particularly to communication protocols. [Tel94], [Bar96], and [Lyn96] in a similar style offer broad collections of algorithms, including temporal aspects such as timing constraints, probabilistic algorithms, etc. In particular, [Lyn96] is an almost complete compendium of distributed algorithms. All these books represent algorithms in pseudo code of I/O automata, and employ semi-formal correctness arguments. In contrast, we consider fewer algorithms, excluding real-time and probabilistic ones. But we suggest an operational model and formal verification techniques that exploit concurrency.

[Bes96] offers a number of algorithms, some of which we consider, too. Among all the textbooks mentioned, this is the only one to model concurrency explicitly (with the help of Petri nets). It also employs a Petri net based technique (transition invariants) to argue about liveness properties. We suggest a version of temporal logic for this purpose.

[MP92] and [MP95] suggest a programming representation for algorithms, together with a formal semantics, focusing on temporal logic-based proof of safety properties. Liveness was postponed to a forthcoming volume. We cover liveness, too.

Summing up, in contrast to our approach, none (but [Bes96]) of the mentioned textbooks employs an operational model that would represent or exploit concurrency explicitly (though concurrency is an essential feature of distributed algorithms). Verification is addressed with a different degree of rigor in all texts, most formally in [CM88], [MP92], [MP95], [Bes96], and [Sch97]. Formal verification always (except in [Bes96]) employs temporal logic on transition systems, thus not exploiting concurrency. In contrast, we suggest a version of temporal logic that exploits concurrency.

2 Basic Definitions: Nets

This section provides the general framework of *state elements*, *transition elements*, and their *combination*. This framework will later be applied in various contexts.

Figure 1.1 shows an example of a net with a particular interpretation: circles and boxes represent *local states* and *actions*, respectively.

There exist other interpretations of nets, too. But they always follow the same scheme: Two sorts of components are identified, emphasizing "passive" and "active" aspects, respectively. They are combined by an abstract relation, always linking elements of *different* sorts.

2.1 Definition. *Let P and T be two disjoint sets, and let $F \subseteq (P \times T) \cup (T \times P)$. Then $N = (P, T, F)$ is called a* net.

Unless interpreted in a special manner, we call the elements of P, T, and F *places*, *transitions*, and *arcs*, respectively. F is sometimes called the *flow relation* of the net.

We employ the usual graphical representation of nets, depicting places, transitions, and arcs as *circles*, *boxes*, and *arrows*, respectively. An arrow $x \rightarrow y$ represents the arc (x, y). Ignoring the black dots inside some of the circles, Fig. 1.1 shows a net with six places and four transitions.

As a matter of convenience, in this text a net will frequently be identified by the number of the figure representing it. As an example, $\Sigma_{1.1}$ denotes the net in Fig. 1.1.

Nets are occasionally denoted in the literature as bipartite graphs. But notice that the two nets of Fig. 2.1 are not equivalent in any relevant context.

The following notational conventions will be employed throughout the entire book:

2.2 Definition. *Let $N = (P, T, F)$ be a net.*

Figure 2.1. Two different nets

i. P_N, T_N, and F_N denote P, T, and F, respectively. By abuse of notation, N often stands for the set $P \cup T$, and aFb for $(a,b) \in F$.

ii. As usual, F^{-1}, F^+, and F^* denote the inverse relation, the transitive closure, and the reflexive and transitive closure of F, respectively, i.e., $aF^{-1}b$ iff bFa, aF^+b iff $aFc_1Fc_2\ldots c_nFb$ for some $c_1,\ldots,c_n \in N$ and aF^*b iff aF^+b or $a = b$. For $a \in N$, let $F(a) = \{b \mid aFb\}$.

iii. Whenever F can be assumed from the context, for $a \in N$ we write $^\bullet a$ instead $F^{-1}(a)$ and $a^\bullet$ instead $F(a)$. This notation is translated to subsets $A \subseteq N$ by $^\bullet A = \bigcup_{a \in A} {}^\bullet a$ and $A^\bullet = \bigcup_{a \in A} a^\bullet$. $^\bullet A$ and $A^\bullet$ are called the pre-set (containing the pre-elements) and the post-set (containing the post-elements) of A.

The following examples for the above notations apply to $\Sigma_{1.1}$ (i.e., the net in Fig. 1.1): For each place $p \in P_{\Sigma_{1.1}}$ both sets $^\bullet p$ and $p^\bullet$ have one element. For each $t \in T_{\Sigma_{1.1}}$, $|{}^\bullet t| = |t^\bullet|$. Furthermore, for all $a, b \in \Sigma_{1.1}$ $a(F^*_{N_{1.1}})b$.

Obviously, for $x, y \in N$, $x \in {}^\bullet y$ iff $y \in x^\bullet$.

The rest of this section introduces basic notions such as *isomorphism*, special substructures of nets, and *subnets*. It may be skipped at first reading.

Isomorphism between nets is defined as can be expected:

2.3 Definition. *Two nets N and N' are* isomorphic *(written: $N \simeq N'$) iff there exists a bijective mapping $\beta : N \to N'$ between their element sets such that $\beta(P_N) = P_{N'}$, $\beta(T_N) = T_{N'}$, and xF_Ny iff $\beta(x)F_{N'}\beta(y)$.*

We are mostly not interested in the individuality of places and transitions of a net. Any isomorphic net does the same job, in general. Nets resemble graphs in this respect. In graphical representations of nets, then, places and transitions remain unnamed. We employed this convention already in notation (1).

The following special structures are frequently distinguished:

2.4 Definition. *Let N be a net.*

i. $x \in N$ *is* isolated *iff* $^\bullet x \cup x^\bullet = \emptyset$.

ii. $x, y \in N$ *form a* loop *iff* xF_Ny *and* yF_Nx

iii. x *and* y *are* detached *iff* $(^\bullet x \cup \{x\} \cup x^\bullet) \cap (^\bullet y \cup \{y\} \cup y^\bullet) = \emptyset$.

iv. *For $A \subseteq N$, N is* A-simple *iff for all $x, y \in A$: $^\bullet x = {}^\bullet y$ and $x^\bullet = y^\bullet$ imply $x = y$.*

v. N *is* simple *iff N is N-simple.*

vi. N *is connected iff for all* $x, y \in N : x(F \cup F^{-1})^* y$.
vii. N *is strongly connected iff for all* $x, y \in N : x(F^*)y$.

As an example, the net $\Sigma_{1.1}$ in Fig. 1.1 has no isolated elements and no loops; it is simple, connected, and even strongly connected. State *buffer filled* and action *produce* are detached, whereas *buffer filled* and *ready to produce* are not. Each of the two nets in Fig. 2.1 is connected, but not strongly connected. Figure 2.2 gives further examples for the special structures described above.

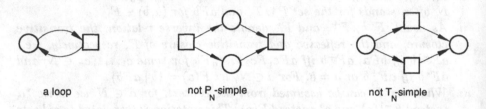

a loop   not P_N-simple   not T_N-simple

Figure 2.2. Special structures in nets

Isolated elements sometimes occur as a technical construct. They have no particularly reasonable use in many applications of nets. Loops occasionally play a distinguished role. Most nets to be studied will be connected. But it is occasionally illuminating to consider two entirely unconnected nets as one net.

Simplicity, as defined in Def. 2.4(iv), is quite a natural assumption or property in a wide range of applications. Each transition t of a T_N-simple net N is uniquely determined by its pre- and postsets $^\bullet t$ and $t^\bullet$. Representing each transition t by $(^\bullet t, t^\bullet)$, N is uniquely given by P_N and T_N. Likewise, each place p of a P_N-simple net N is uniquely determined by $^\bullet p$ and $p^\bullet$.

To sum up the potential links between two elements of a net, Def. 2.1 implies that elements of equal type (i.e., two places or two transitions) are never F-related. Each pair of elements of different type correspond in exactly one out of four alternative ways, as shown in Fig. 2.3.

Nets are frequently used in a *labeled* version, with some symbols or items assigned to places, transitions, or arcs.

2.5 Definition. *Let N be a net and let A be any set.*

i. *Let* $l_1 \colon P_N \to A$, $l_2 \colon T_N \to A$, $l_3 \colon P_N \cup T_N \to A$ *and* $l_4 \colon F_N \to A$ *be mappings.* $l_1, \ldots, l_4$ *are called a* place labeling, *a* transition labeling, *an* element labeling, *and an* arc labeling *of N over A, respectively.*
ii. *N is said to be* place labeled (transition labeled, element labeled, arc labeled, *respectively*) *over A iff a corresponding labeling is given either explicitly or implicitly from the context.*

Labelings are graphically represented by means of symbols ascribed to the corresponding circles, boxes, or arrows. For example, the dots in some circles

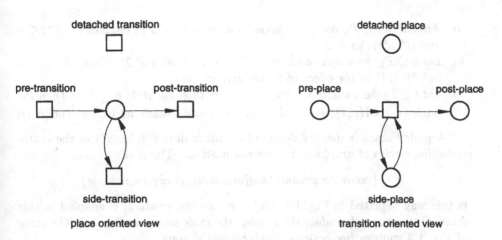

Figure 2.3. The relationship between places and transitions

of Figs. 1.1 – 1.3 represent a place labeling $l : P_\Sigma \to \{0,1\}$, with $l(p) = 1$ iff the circle representing p carries a dot.

An already labeled net may get additional labelings.

3 Dynamics

Figure 1.1 shows a net with places and transitions interpreted as *local states* and *actions*, respectively. A set of local states forms a *global state*. Its elements are graphically depicted by a dot in the corresponding circle. Section 1.6 explained that an action t is about to *occur* in a global state, provided $^\bullet t$ belongs to that state. Occurrence of t then replaces $^\bullet t$ by $t^\bullet$, this way yielding a new state, as graphically shown in Fig. 3.1.

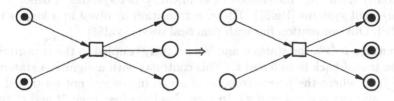

Figure 3.1. Occurrence of an action

Nets will be denoted by Σ; places and transitions will be called *local states* and *actions*, respectively, to underline this interpretation.

3.1 Definition. *Let Σ be a net.*

i. *Any subset $a \subseteq P_\Sigma$ of local states is called a* (global) *state of Σ.*

    ii. An action $t \in T_\Sigma$ has concession in a given state a (a enables t) iff $^\bullet t \subseteq a$ and $(t^\bullet \setminus {}^\bullet t) \cap a = \emptyset$.

   iii. Let $a \subseteq P_\Sigma$ be a state and let $t \in T_\Sigma$ be an action of Σ. Then $\mathrm{eff}(a,t) := (a \setminus {}^\bullet t) \cup t^\bullet$ is the effect of t's occurrence on a.

   iv. Let $t \in T_\Sigma$ be an action with concession in some state $a \subseteq P_\Sigma$. Then the triple $(a, t, \mathrm{eff}(a,t))$ is called a step in Σ and usually written $a \xrightarrow{t} \mathrm{eff}(a,t)$.

A global state is usually depicted by black dots ("tokens") in the corresponding circles of graphical net representations. The state

$$a = \{ ready\_to\_produce, buffer\_empty, ready\_to\_remove \}$$

is this way depicted in Fig. 1.1. Only one action, *produce*, is enabled in this state. Occurrence of *produce* then yields the state shown in Fig. 1.2. The state of Fig. 1.3 enables *two* actions, *produce*, and *remove*.

    Intuitively, $^\bullet t$ is the set of *pre-conditions* for the occurrence of action t, and $t^\bullet$ is the set of conditions holding after t's occurrence (we may call them *post-conditions* of t).

    The above definition invites a number of observations, to be discussed in the rest of this section. First of all, a transition involved in a loop, as in Fig. 3.2, may very well have concession in some given state a. This deviates

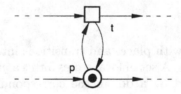

Figure 3.2. A loop

essentially from the conventions of elementary net systems [Roz86] or condition/event systems [Rei85]. There, a transition involved in a loop is never enabled. Our convention fits with practical needs [Val86].

    In a step $a \xrightarrow{t} b$, the states a and b are tightly coupled to the transition t: a can be traced back from b and t. (This contrasts with assignment statements $x := f(x)$, where the previous value of x can in general not be traced back from f and the new value of x). In case Σ is loop-free, even $^\bullet t$ and $t^\bullet$ can be retrieved from a and b.

3.2 Lemma. *Let $a \xrightarrow{t} b$ be a step of some net Σ.*

   i. $a = (b \setminus t^\bullet) \cup {}^\bullet t$.

  ii. $^\bullet t = a \setminus b$ and $t^\bullet = b \setminus a$ iff Σ is loop-free.

In a step $a \xrightarrow{t} b$, a set c of places may be added or be removed from both a and b, provided c is disjoint from $^\bullet t$ and from $t^\bullet$:

3.3 Lemma. *Let* $a \xrightarrow{t} b$ *be a step of some net* Σ *and let* $c \subseteq P_\Sigma$ *with* $c \cap (^\bullet t \cup t^\bullet) = \emptyset$.

   i. $(a \cup c) \xrightarrow{t} (b \cup c)$ is a step of Σ.

   ii. $(a \setminus c) \xrightarrow{t} (b \setminus c)$ is a step of Σ.

We leave proof of Lemmas 3.2 and 3.3 as an exercise for the reader. Generally, steps exhibit a whole bunch of symmetries, particularly for loop-free es-nets.

Two situations deserve particular attention: Firstly we observe that according to Def. 3.1(ii) there can be two reasons for a transition t not to have concession in some state a: either some precondition is missing ($^\bullet t \not\subseteq a$), or they are all present ($^\bullet t \subseteq a$), but additionally one of the "new" postconditions is already present ($(t^\bullet \setminus {}^\bullet t) \cap a \neq \emptyset$), as in Fig. 3.3. This kind of situation will be denoted as *contact*:

3.4 Definition. *Let Σ be a net with a transition $t \in T_\Sigma$ and u state $a \subseteq P_\Sigma$. Then a is a* contact state *with respect to t iff $^\bullet t \subseteq a$ and $(t^\bullet \setminus {}^\bullet t) \cap a \neq \emptyset$.*

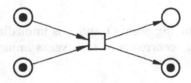

Figure 3.3. Contact situation

Hence, in case of no contact, the preconditions alone provide the requirements of enabling.

The second situation concerns two transitions t and u, both enabled in some state a. If they share a common pre- or postcondition, as in Fig. 3.4, the occurrence of t prevents the occurrence of u (and vice versa); t and u are then said to be in *conflict*.

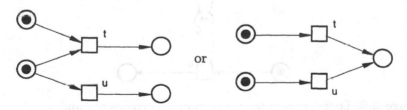

Figure 3.4. Conflict situation

3.5 Definition. *Let Σ be a net, with two different transitions $t, u \in T_\Sigma$ and a state $a \subseteq P_\Sigma$: a is a conflict state with respect to t and u iff both t and u have concession in a, and are not detached.*

The state of the net shown in Fig. 3.5 is conflicting with respect to a and b, as well as with respect to a and c. The two actions b and c are not conflicting.

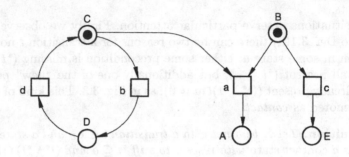

Figure 3.5. Net with conflict

The notion of conflicting events t and u is immediately obvious for loop-free nets. In this case, occurrence of t prevents immediate occurrence of u (and vice versa).

3.6 Lemma. *Let Σ be a loop-free net. Let a be a conflict state with respect to two transitions t and u of Σ. Then $a \xrightarrow{t} b$ implies u not be enabled in state b.*

In the context of loops, as in Fig. 3.6, conflict between t and u prevents t and u occurring *concurrently*. A formal definition of events occurring concurrently is postponed to Sect. 5.

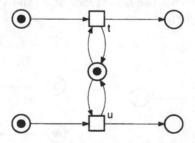

Figure 3.6. Loops, preventing concurrent occurrence of t and u

4   Interleaved Runs

Single steps of a net Σ, as considered in the previous section, compose to *runs* of Σ. The most elementary composition of two steps $a \xrightarrow{t} b$ and $b \xrightarrow{u} c$ is their sequential combination in the run $a \xrightarrow{t} b \xrightarrow{u} c$. Generally, one may construct runs $a_0 \xrightarrow{t_1} a_1 \xrightarrow{t_2} \cdots \xrightarrow{t_n} a_n$ of nets Σ, provided $a_{i-1} \xrightarrow{t_i} a_i$ is a step of Σ, for $i = 1, \ldots, n$. Furthermore, we consider infinite runs, too:

4.1 Definition. *Let Σ be a net.*

  i. *For $i = 1, \ldots, n$ let $a_{i-1} \xrightarrow{t_i} a_i$ be steps of Σ. Those steps form a Σ-based finite interleaved run w, written $a_0 \xrightarrow{t_1} a_1 \xrightarrow{t_2} \cdots \xrightarrow{t_n} a_n$. Each $i \in \{0, \ldots, n\}$ is an index of w.*

 ii. *For each $i = 1, 2, \ldots$ let $a_{i-1} \xrightarrow{t_i} a_i$ be steps of Σ. Those steps form a Σ-based infinite interleaved run w, sometimes outlined $a_0 \xrightarrow{t_1} a_1 \xrightarrow{t_2} \cdots$. Each $i \in \mathbb{N}$ is an index of w.*

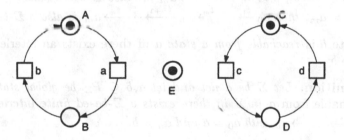

Figure 4.1. A net consisting of independent subnets

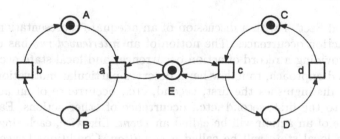

Figure 4.2. Extending $\Sigma_{4.1}$ by loops

Examples of finite runs of the net $\Sigma_{1.1}$ include $a_0 \xrightarrow{produce} a_1 \xrightarrow{deliver} a_2 \xrightarrow{remove} a_3 \xrightarrow{produce} a_4$ and $a_0 \xrightarrow{produce} a_1 \xrightarrow{deliver} a_2 \xrightarrow{produce} a_5 \xrightarrow{remove} a_4$, with a_0 as depicted in Fig. 1.1 and $a_1, \ldots, a_5$ obvious from context. Each finite run of $\Sigma_{1.1}$ can be extended to infinitely many finite and infinite runs. Figures 4.1 and 4.2 show two different nets. The interleaved runs (both finite and infinite) starting at the depicted global state are equal for both nets.

The runs of a net exhibit some regularities. First we consider runs consisting of two steps. We give a Lemma for loop-free nets and leave the general case to the reader:

4.2 Lemma. *Let Σ be a loop-free net, and let $a \xrightarrow{t} b \xrightarrow{u} c$ be a Σ-based run.*

i. ${}^\bullet t \cap {}^\bullet u = t^\bullet \cap u^\bullet = \emptyset$.

ii. There exists a state d with $a \xrightarrow{u} d \xrightarrow{t} c$ iff t and u are detached.

Each initial part of a run is a run. Furthermore, "cyclic" sequences of steps can be repeated:

4.3 Lemma. *Let Σ be a net, let $a_0 \xrightarrow{t_1} \cdots \xrightarrow{t_m} a_m$ be a Σ-based run, and let $n < m$.*

i. $a_0 \xrightarrow{t_1} \cdots \xrightarrow{t_n} a_n$ and $a_n \xrightarrow{t_n} \cdots \xrightarrow{t_m} a_m$ are also Σ-based runs.

ii. If $a_n = a_m$, then $a_0 \xrightarrow{t_1} \cdots \xrightarrow{t_m} a_m \xrightarrow{t_{n+1}} \cdots \xrightarrow{t_m} a_m$ is also a Σ-based run.

A state b is *reachable from* a state a iff there exists an interleaved run from a to b:

4.4 Definition. *Let Σ be a net and let $a, b \subseteq P_\Sigma$ be global states of Σ: b is* reachable from a *in Σ iff there exists a Σ-based finite interleaved run $a_0 \xrightarrow{t_1} a_1 \xrightarrow{t_2} \ldots \xrightarrow{t_n} a_n$ with $a_0 = a$ and $a_n = b$.*

We leave proof of Lemmas 4.2 and 4.3 as an exercise for the reader.

5   Concurrent Runs

We finished Sect. 4 with a discussion of an adequate representation of independent action occurrences. The notion of an *interleaved run* has been suggested, providing a record of action occurrences and local state occurrences. The revised approach, to follow here, deserves particular motivation.

A run distinguishes the first, second, etc., occurrence of an action and relates it to the first, second ,etc., occurrence of other actions. Each single occurrence of an action will be called an *event*. Likewise, each single occurrence of a local state will be called a *condition*. Conditions hence serve as *preconditions* and *postconditions* for events. Then, a run consists of conditions and events, ordered by a "before – after" relation. Interleaved runs, discussed in Sect. 4, provide global states and an order on events, motivated by an "observer" who observes events one after the other. Different observers may observe different orders of events, hence a net is associated with a *set* of interleaved runs.

This concept confuses system-specified, *causal* order with order additionally introduced by *observation*. Events that occur independently are arbitrarily ordered by observation. Even if we assume that independence among events may not be observable, it may nevertheless be representable. So we ask for a representation of objective, i.e., entirely system-based, ordering of conditions and events.

Before formally defining such a notion, we discuss some of the properties to be expected from this concept.

Firstly, independent events should be distinguished from events in arbitrary order. As an example, compare $\Sigma_{4.1}$ and $\Sigma_{4.2}$: a and c occur independently in $\Sigma_{4.1}$, whereas in $\Sigma_{4.2}$ they occur in either order.

The essential difference between $\Sigma_{4.1}$ and $\Sigma_{4.2}$ is the existence of *conflict* in $\Sigma_{4.2}$: Whenever the state shown in Fig. 4.2 has been reached, a decision has to be made concerning the order of a's and c's occurrence. Different outcomes of this decision yield different runs. Hence $\Sigma_{4.2}$ evolves different runs, in fact infinitely many different runs (because the state of $\Sigma_{4.2}$, shown in Fig. 4.2, is reached infinitely often).

A state in $\Sigma_{4.1}$ never occurs where a decision between enabled actions is to be made: Whenever two actions are enabled, they occur mutually independently.

To sum up, an observer-independent notion of runs should record events and conditions. It should make explicit to what extent events and conditions are ordered due to the underlying system's constraints. Hence, this kind of occurrence record partially orders its elements by the relation "x is a causal prerequisite for y", because repetitions of the same action or the same local state are recorded as new entries. Unordered elements denote independent ("concurrent") occurrences. There is a fairly obvious representation of such records, namely again as a net. Figures 5.1 – 5.5 show examples.

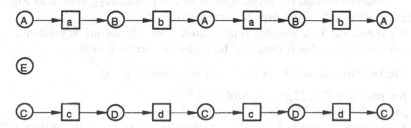

Figure 5.1. Concurrent run of $\Sigma_{4.1}$

Each transition in Figs. 5.1 and 5.2 represents an event, i.e., the occurrence of an action. This action is denoted by the transition's labeling. Distinct transitions with the same labeling denote different occurrences of the same action. Similarly, a place q shows by its inscription b that local state b has been reached due to the occurrence of $^\bullet q$ and has been left as a result of the occurrence of $q^\bullet$.

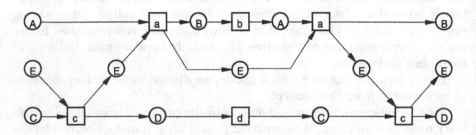

Figure 5.2. Concurrent run of $\Sigma_{4.2}$

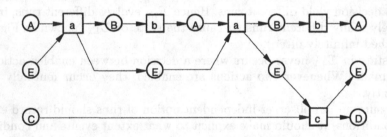

Figure 5.3. Concurrent run of $\Sigma_{4.2}$

Figure 5.1 shows that the behavior of $\Sigma_{4.1}$ consists of two independent sequences. Figure 5.2 likewise shows a concurrent run of $\Sigma_{4.2}$, where the first occurrence of c is before the first occurrence of a, and the second occurrence of c is after the second occurrence of a. In the run of $\Sigma_{4.2}$ shown in Fig. 5.3, a occurs twice before the first occurrence of c.

A *concurrent run* will be represented formally as an acyclic net with unbranched places. Such nets will be called *occurrence nets*.

5.1 Definition. *A net K is called an* occurrence net *iff*

  i. *for each $p \in P_K$, $|^\bullet p| \leq 1$ and $|p^\bullet| \leq 1$,*
 ii. *for each $t \in T_K$, $|^\bullet t| \geq 1$ and $|t^\bullet| \geq 1$,*
iii. *the transitive closure F_K^+ of F_K, frequently written $<_K$, is irreflexive (i.e.,*
    $x_1 F_K x_2 F_K \ldots F_K x_n$ *implies* $x_1 \neq x_n$),
 iv. *for each $x \in K$, $\{y \mid y <_K x\}$ is finite.*

Figures 5.1–5.3 show labeled occurrence nets. $<_K$ is a strict partial order in each occurrence net K. In fact, $x <_K y$ iff there exists an arrow sequence from x to y.

We are particularly interested in states consisting of pairwise unordered places:

5.2 Definition. *Let K be an occurrence net.*

    i. Two elements $p, q \in K$ are concurrent iff neither $p <_K q$ nor $q <_K p$.
    ii. A state $a \subseteq P_K$ is concurrent iff its elements are pairwise concurrent.
    iii. A state a is maximal concurrent *iff a is concurrent and no $p \in a$ is concurrent to any $q \in K \setminus a$.*
    iv. Let $^{\circ}K := \{k \in K \mid {}^{\bullet}k = \emptyset\}$ and let $K^{\circ} := \{k \in K \mid k^{\bullet} = \emptyset\}$.

Occurrence of actions preserves concurrency:

5.3 Lemma. *Let K be an occurrence net and let $a \xrightarrow{t} b$ be a step of K.*

    i. If a is concurrent, then b is concurrent, too.
    ii. If a is maximal concurrent, then b is maximal concurrent, too.

Proof of this lemma is left as an exercise for the reader.

According to the intended use (described above) of an occurrence net K to describe a run of a net Σ, each maximal concurrent state a of K represents a state of Σ that might have been observed during the course of K. Two a-enabled actions of K represent concurrent (independent) occurrences of the corresponding actions of Σ.

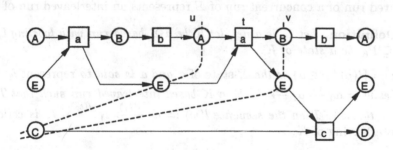

Figure 5.4. A step of a concurrent run of $\Sigma_{4.2}$

5.4 Definition. *Let Σ be a net, let K be an occurrence net and let $l : K \to \Sigma$ be an element labeling of K. K is a Σ-based* concurrent run *iff*

    i. concurrent elements of K are differently labeled,
    ii. for each $t \in T_K$, $l(t) \in T_{\Sigma}$, $l({}^{\bullet}t) = {}^{\bullet}l(t)$ and $l(t^{\bullet}) = l(t)^{\bullet}$.

According to this definition, Fig. 5.1 in fact shows a $\Sigma_{4.1}$-based concurrent run. Figures 5.2 and 5.3 likewise show $\Sigma_{4.2}$-based runs. A step $u \xrightarrow{t} v$ is additionally outlined in Fig. 5.4. With l denoting the labeling of Fig. 5.4, $l(u) \xrightarrow{l(t)} l(v)$ is the step $\{A, C, E\} \xrightarrow{a} \{B, C, E\}$ of $\Sigma_{4.2}$. Figure 5.5 shows a

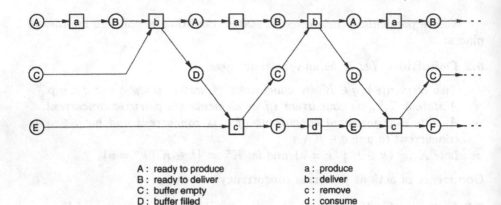

A : ready to produce   a : produce
B : ready to deliver   b : deliver
C : buffer empty   c : remove
D : buffer filled   d : consume
E : ready to remove
F : ready to consume

Figure 5.5. The unique maximal concurrent run of $\Sigma_{1.1}$

further example. Just like $\Sigma_{4.1}$, and in contrast to $\Sigma_{4.2}$, the net $\Sigma_{1.1}$ evolves a unique maximal concurrent run.

The above definition meets the intuition of concurrent runs only as long as no contact states occur (cf. Def. 3.4). We stick to such runs in the sequel.

Interleaved and concurrent runs of a net Σ are tightly related: Each interleaved run of a concurrent run of Σ represents an interleaved run of Σ.

5.5 Definition. *Let Σ be a net, let K be a Σ-based run with labeling l, and let $a \subseteq P_K$ be a state of K.*

i. *$\hat{a} := \{l(p) \mid p \in a\}$ is the Σ-state of a and a is said to* represent *$\hat{a}$.*

ii. *Let $w = a_0 \xrightarrow{t_1} a_1 \xrightarrow{t_2} \ldots$ be a K-based interleaved run such that $T_K = \{t_1, t_2, \ldots\}$. Then the sequence $l(w) := \hat{a}_0 \xrightarrow{l(t_1)} \hat{a}_1 \xrightarrow{l(t_2)} \ldots$ is called an* interleaving *of K.*

5.6 Lemma. *Let Σ be a net.*

i. *Let K be a Σ-based concurrent run. Then each interleaving of K is a Σ-based interleaved run.*

ii. *Let v be a Σ-based interleaved run. Then there exists a unique Σ-based concurrent run K such that v is an interleaving of K.*

Proof of this lemma is left as an exercise for the reader.

Writing sets $\{X, Y, Z\}$ as XYZ, the following are two examples of interleaved runs of $\Sigma_{4.2}$:

$$v_1 = ACE \xrightarrow{c} ADE \xrightarrow{a} BDE \xrightarrow{b} ADE \xrightarrow{d} ACE \xrightarrow{a} BCE \xrightarrow{c} BDE, \tag{1}$$

$$v_2 = ACE \xrightarrow{c} ADE \xrightarrow{a} BDE \xrightarrow{d} BCE \xrightarrow{b} ACE \xrightarrow{a} BCE \xrightarrow{c} BDE. \tag{2}$$

There exists two interleaved runs w_1 and w_2 of the run of $\Sigma_{4.2}$ given in Fig. 5.2 such that $v_1 = l(w_1)$ and $v_2 = l(w_2)$.

Hence the concurrent runs of a net Σ partition the set of interleaved runs of Σ into equivalence classes, where two interleaved runs v_1 and v_2 are equivalent iff there exists a concurrent run K of Σ with two interleaved runs w_1 and w_2 such that $l(w_1) = v_1$ and $l(w_2) = v_2$.

6 Progress

Any description of algorithms usually goes with the implicit assumption of *progress*. As an example, each execution of a PASCAL program is assumed to continue as long as the program counter points at some executable statement; intermediate termination at some executable statement is not taken into account. The situation is more involved for distributed algorithms. Progress is usually assumed for *most*, but not necessarily all actions.

As an example, one may intend $\Sigma_{1.1}$ not to terminate in a state s with $\{ready\ to\ deliver, empty\} \subseteq s$, i.e., with *deliver* enabled. Likewise one may want *receive* and *consume* not to remain enabled indefinitely. Not enforcing *produce* may be adequate, however; this action may depend on components not represented in Fig. 1.1. So one may be interested in runs that may *neglect* progress of *produce*, but *respect* progress of all other actions.

6.1 Definition. *Let Σ be a net and let $t \in T_\Sigma$.*

i. *A Σ-based finite or infinite interleaved run $w = a_0 \xrightarrow{t_1} a_1 \xrightarrow{t_2} \ldots$ neglects progress of t iff some state a_i enables t, and for no index $j > i$, $t_j \in (^\bullet t)^\bullet$.*
ii. *A Σ-based concurrent run K with labeling l neglects progress of t iff $l(K^\circ)$ enables t.*
iii. *An interleaved or concurrent run r respects progress of t iff r does not neglect progress of t.*

The concurrent run in Fig. 5.1 respects progress of b and d, and neglects progress of a and c. The infinite run outlined in Fig. 5.5 respects progress of all actions of $\Sigma_{1.1}$ A run r of the conflicting net $\Sigma_{3.5}$ respects progress of all its actions if a is the last action to occur in r, or if d and b occur infinitely often and c just once in r.

Progress is sensitive to loops. For example, Fig. 6.1 shows a net consisting of two detached parts, and Fig. 6.2 gives a $\Sigma_{6.1}$-based concurrent run K. This run obviously neglects progress of c in K, because K can be extended, as in Fig. 6.3.

The run K has a unique interleaving

$$w = \{A, C\} \xrightarrow{a} \{B, C\} \xrightarrow{b} \{A, C\} \xrightarrow{a} \ldots \tag{1}$$

which likewise neglects progress of c in $\Sigma_{6.1}$.

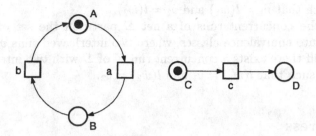

Figure 6.1. A net consisting of two detached parts

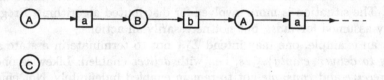

Figure 6.2. $\Sigma_{6.1}$-based concurrent run, neglecting progress of c

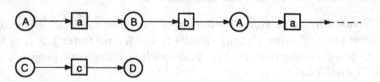

Figure 6.3. $\Sigma_{6.1}$-based concurrent run, respecting progress of c

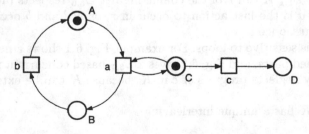

Figure 6.4. The door control system

Figure 6.4 now extends $\Sigma_{6.1}$ by a loop (a, C), and Fig. 6.5 gives a $\Sigma_{6.4}$-based concurrent run, K'. This run respects progress of c very well. Unlike the run K of Fig. 6.2, the run K' can not be extended by an occurrence of c, because C is indefinitely engaged in the occurrence of a. Just like K, the run K' has a unique interleaving; furthermore, it is exactly the same interleaving as K, given in (1). Each state of (1) is followed by an occurrence of action a. This action conflicts with c in $\Sigma_{6.4}$ ($a \in (^\bullet c)^\bullet$), hence (1) respects progress of c in $\Sigma_{6.4}$.

To sum up, occurrence of progress respecting action t in an interleaved run w is not guaranteed by its *persistent* enabling (i.e., enabling in each state of w, as of c in (1)), but only by its persistent *and conflict free* enabling.

The following interpretation of $\Sigma_{6.4}$ shows that this conduct of progress perfectly matches intuition: Assume a crowd of people, occasionally passing a gate (action a). Local state A is taken whenever a person is due to pass the gate. Passage is feasible only in case the gate is not locked (state C). Furthermore, a guard is supposed to lock the gate (action c). Locking and passing the gate (actions a and c) are conflicting actions. Progress of a and c just ensures that either of them will occur in the state shown in Fig. 6.4.

The run in Fig. 6.5 shows the case of continuous heavy traffic at the gate, "preventing" the guard from closing the gate.

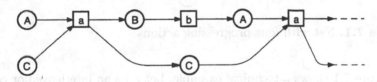

Figure 6.5. $\Sigma_{6.4}$-based concurrent run, respecting progress of c

Defs. 6.1(i) and 6.1(ii) of progress are closely related: K respects progress of t iff each interleaving of K does:

6.2 Lemma. *Let Σ be a net, let K be a Σ-based concurrent run and let $t \in T_\Sigma$. Then K respects progress of t iff each interleaving of K respects progress of t.*

Proof of this lemma is left as an exercise for the reader.

The assumption of progress resembles the well known assumption of weak fairness for some action t. This assumption rules out an interleaved run $w = a_0 \xrightarrow{t_1} a_1 \xrightarrow{t_2} \ldots$ where for some $n \in \mathbb{N}$ all states a_{n+i} enable t, but no t_{n+i} is equal to t.

Progress and weak fairness coincide for the case of loop-free systems. The above example, however, shows a subtle difference in the case of loops: The interleaved run w of (1) is not weakly fair for action c in the net $\Sigma_{6.4}$, but w

very well respects progress of c in $\Sigma_{6.4}$. Conversely, each progress respecting interleaved run is weakly fair.

7 Fairness

Many distributed algorithms require the assumption of *fairness* for some actions. Intuitively formulated, a single run r *neglects* fairness of some action t iff t occurs only finitely often, but is enabled infinitely often in r. Such runs will be discarded in case fairness is assumed for t.

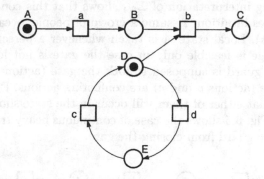

Figure 7.1. Net with four progressing actions

Figure 7.1 shows a technical example. Let r be an interleaved or concurrent run of $\Sigma_{7.1}$, respecting progress of all actions. Then a occurs and b is eventually enabled in r. Either b eventually occurs in r, or b is infinitely often enabled in r. In the latter case, r neglects fairness for b.

7.1 Definition. *Let Σ be a net and let $t \in T_\Sigma$.*

i. *A Σ-based interleaved run w neglects fairness for t iff t occurs only finitely often in w and is enabled infinitely often in w.*

ii. *A Σ-based interleaved run w respects fairness for t iff w does not neglect fairness for t.*

iii. *A Σ-based concurrent run K respects fairness of t iff all interleavings of K respect fairness of t.*

An example is the infinite interleaved run of $\Sigma_{7.1}$:

$$AD \xrightarrow{a} BD \xrightarrow{d} BE \xrightarrow{c} BD \xrightarrow{d} \ldots \tag{1}$$

Action b is enabled in each occurrence of BD, hence infinitely often. Furthermore, b never occurs in (1), hence (1) neglects fairness for b. Likewise, the concurrent run $K =$

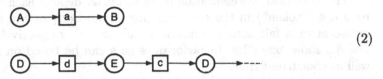

$$(2)$$

of $\Sigma_{7.1}$ neglects fairness of b: The above run (1) is an interleaving of K. Each finite prefix of (1) or (2) respects fairness of all involved actions, but neglects progress of some action.

As a further example, the run shown in Fig. 6.5 of $\Sigma_{6.4}$, though respecting progress for c, does neglect fairness for c.

8 Elementary System Nets

The previous chapters provided all means to model a great variety of distributed algorithms; in fact all algorithms which have a fixed topology, and are governed by control rather than by values. Those means include local and global states, actions and their occurrence, interleaved and concurrent runs, assumptions of progress and quiescence, and fairness. A net that takes into account all such aspects and additionally fixes a distinguished *initial* state, is called an *elementary system net*:

8.1 Definition. *A net Σ is called an* elementary system net *(es-net, for short) iff*

  i. a state $a_\Sigma \subseteq P_\Sigma$ is distinguished, called the initial state *of Σ,*
  ii. each action $t \in T_\Sigma$ is denoted as either progressing or quiescent,
  iii. some progressing actions may be distinguished as fair.

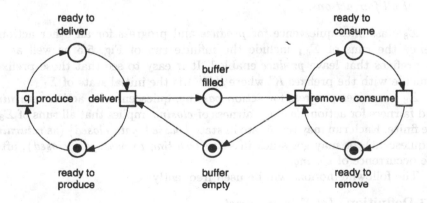

Figure 8.1. Producer/consumer system, assuming quiescence for *produce* and progress for all other actions

The graphical representation of an es-net Σ depicts each element of a_Σ by a dot ("token") in the corresponding circle. Each square representing a quiescent or a fair action is inscribed "q" or "φ", respectively. Figures 8.1 and 8.2 show examples. Behavior of es-nets can be based on interleaved as well as concurrent runs.

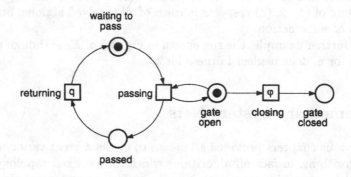

Figure 8.2. The Door control system, assuming progress for *passing*, quiescence for *returning* and fairness for *closing*

8.2 Definition. *Let Σ be an es-net.*

  i. A Σ-based interleaved run $w = a_0 \xrightarrow{t_1} a_1 \xrightarrow{t_2} \ldots$ is an interleaved run *of Σ iff $a_0 = a_\Sigma$ and w respects progress of all progressing actions and fairness of all fair actions of Σ.*

  ii. A Σ-based concurrent run K with labeling l is a concurrent run *of Σ iff $l(^\circ K) = a_\Sigma$, K respects progress of all progressing actions and fairness of all fair actions.*

$\Sigma_{8.1}$ assumes quiescence for *produce* and progress for all other actions. Hence the runs of $\Sigma_{8.1}$ include the infinite run of Fig. 5.5 as well as all its prefixes that leave *produce* enabled. It is easy to see that those prefixes coincide with the prefixes K' where $l(K'^\circ)$ is the initial state of $\Sigma_{1.1}$.

$\Sigma_{8.2}$ assumes progress for action *passing*, quiescence for action *returning* and fairness for action *closing*. Fairness of *closing* implies that all runs of $\Sigma_{8.2}$ are finite. Each run may terminate in state $\{passed, gate\ closed\}$ (as *returning* is quiescent), or may get stuck in state $\{waiting\ to\ pass, gate\ closed\}$, after the occurrence of *closing*.

The following notions will be used frequently:

8.3 Definition. *Let Σ be an es-net.*

  i. $a \subseteq P_\Sigma$ is a reachable state *of Σ iff a is reachable from a_Σ.*

  ii. $t \in T_\Sigma$ is a reachable action *iff t is enabled in some reachable state.*

iii. Σ *is* conflict free *iff no reachable state is a conflict state.*
iv. Σ *is* contact free *iff no reachable state is a contact state.*

Es-nets considered in the sequel will usually be contact free.

8.4 Lemma. *Let Σ be an es-net without quiescent actions. Σ is conflict free iff there exists exactly one concurrent run of Σ.*

Proof of this lemma is left as an exercise for the reader.

II. Case Studies

The elementary concepts introduced in Chap. I suffice to adequately model
a broad choice of distributed algorithms. Such algorithms typically stick to
control flow of concurrent systems. Data dependent algorithms will follow in
Part B of this book.

We concentrate on *modeling* here. Formulation and proof of properties
will remain on an intuitive footing.

9   Sequential and Parallel Buffers

This case study extends the producer/consumer system of Fig. 8.1, extending
its one-item buffer to two cells. This can be done in *sequential* and in *parallel*
variants. $\Sigma_{9.1}$ gives the sequential solution: Two buffer cells are arranged
one after the other. A parallel solution is given with $\Sigma_{9.2}$. Being ready to

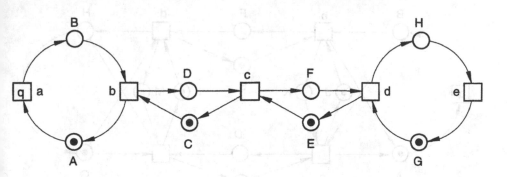

A : ready to produce	E : second buffer cell empty
B : ready to deliver	F : second buffer cell filled
C : first buffer cell empty	G : ready to remove
D : first buffer cell filled	H : ready to consume
a : produce	d : remove
b : deliver	e : consume

Figure 9.1. Producer/consumer with sequential buffer cells

deliver (B), the producer may choose either of the two buffer cells (if both are empty). If one or both are still filled, the producer may employ the empty one or the one that gets empty next, respectively.

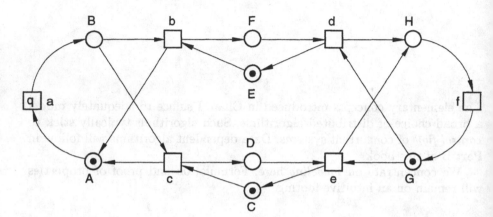

Figure 9.2. Nondeterministic producer/consumer with parallel buffer cells

$\Sigma_{9.2}$ is intuitively "more concurrent" than $\Sigma_{9.1}$ (a notion which will be made more precise later). But "overtaking" is possible in $\Sigma_{9.2}$. As an example, the first buffer cell may be filled before, but emptied after the second one.

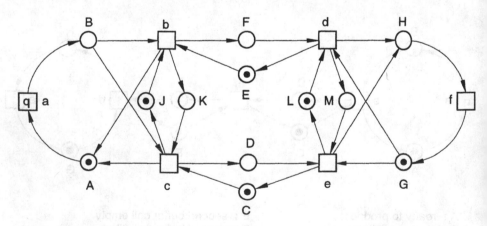

Figure 9.3. Deterministic producer/consumer with parallel buffer cells

Can the advantages of $\Sigma_{9.1}$ (no overtaking) and of $\Sigma_{9.2}$ (direct access to empty buffer cells) be combined? $\Sigma_{9.3}$ shows that this is in fact possible: Access to the buffer cells is organized alternately. But it remains to be shown that $\Sigma_{9.3}$ is "optimal" in some sense: The producer is never given access to a filled buffer cell while the other cell is empty. Nor is the consumer ever given access to an empty buffer cell while the other one is filled.

Unique, formal description of such properties, as well as proof of their correctness, are subject to Part C.

Some differences among $\Sigma_{9.1}$, $\Sigma_{9.2}$, and $\Sigma_{9.3}$ can be studied with the help of their runs: $\Sigma_{9.1}$ has exactly one maximal run (up to isomorphism, cf. Sect. 2). Figure 9.4 shows an initial part of this (periodically structured) run. $\Sigma_{9.3}$ has likewise exactly one maximal run, shown in Fig. 9.6.

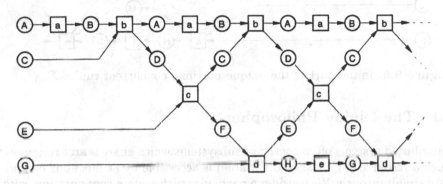

Figure 9.4. Initial part of the unique maximal concurrent run of $\Sigma_{9.1}$

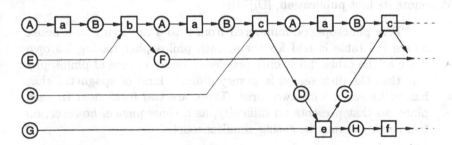

Figure 9.5. A concurrent run of $\Sigma_{9.2}$

Hence, both $\Sigma_{9.1}$ and $\Sigma_{9.3}$ are deterministic (c.f. Lemma 8.4). In contrast, the net $\Sigma_{9.2}$ has infinitely many different maximal runs: Whenever condition B holds, there is a choice between b and c. One of the runs of $\Sigma_{9.2}$ can be gained from $\Sigma_{9.3}$'s run in Fig. 9.6 by skipping all occurrences of the conditions J, K, L and M. A further, extremely "unfair" one is given in Fig. 9.5: the first buffer cell is initially filled, but never emptied.

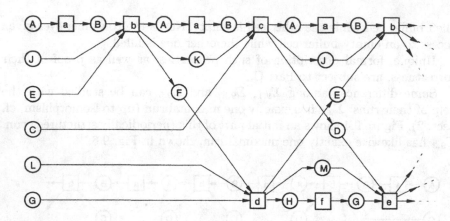

Figure 9.6. Initial part of the unique maximal concurrent run of $\Sigma_{9.3}$

10   The Dining Philosophers

Distributed systems often consist of subsystems which share scarce resources. Such a resource (e.g., a shared variable) is accessible by at most one component simultaneously. We consider a particular such system configuration, with each resource shared by two subsystems, and each subsystem simultaneously requiring two resources. E. W. Dijkstra illustrated this system by "philosophers" and "forks" which stand for subsystems and resources, respectively. We quote its first publication, [Dij71]:

> "Five philosophers, numbered from 0 to 4 are living in a house where the table is laid for them, each philosopher having his own place at the table. Their only problem – besides those of philosophy – is that the dish served is a very difficult kind of spaghetti, that has to be eaten with two forks. There are two forks next to each plate, so that presents no difficulty, as a consequence, however, no two neighbors may be eating simultaneously."

Our first goal is a representation of this system as an es-net, such that the runs of the net describe the "meals" of the philosophers' dinner party.

Figure 10.1 shows this es-net. The philosophers are denoted $A, \dots, E$. Indices p, r, t, e stand for "picks up forks", "returns forks", "thinking", and "eating", respectively. For $i = 0, \dots, 4$, condition a_i denotes that fork i is available for its users.

Each philosopher may start eating in the initial state. But neighboring philosophers apparently compete for their shared fork.

A typical interleaved run of $\Sigma_{10.1}$ is

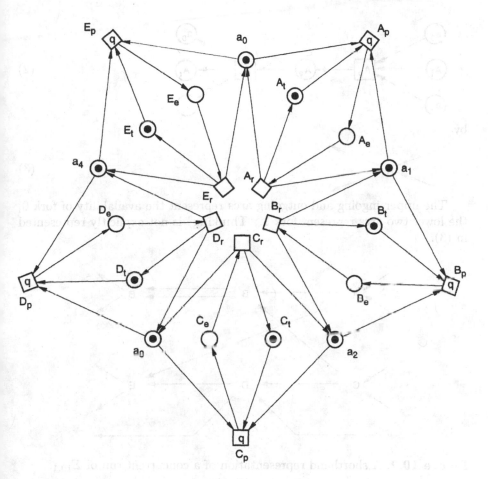

Figure 10.1. The five dining philosophers

$$a_\Sigma \xrightarrow{A_p} a_1 \xrightarrow{C_p} a_2 \xrightarrow{C_r} a_3 \xrightarrow{A_r} a_4 \xrightarrow{B_p} a_5 \xrightarrow{D_p} a_6 \xrightarrow{B_r} a_7 \xrightarrow{B_p} a_8$$
$$\xrightarrow{D_r} a_9 \xrightarrow{E_p} a_{10} \xrightarrow{E_r} a_{11} \xrightarrow{B_r} a_{12} \tag{1}$$

with states $a_1, \dots, a_{12}$ obvious from the context. Philosopher B eats twice in this run, and every other philosopher just once. The final state, a_{12}, coincides with a_Σ.

Turning now to concurrent runs, it is convenient to introduce a shorthand representation for pieces of runs. For philosopher A, call any occurrence of an *eating cycle of A*. We represent each eating cycle of philosopher A

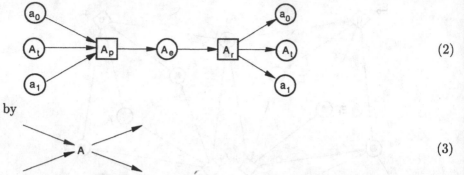

(2)

by

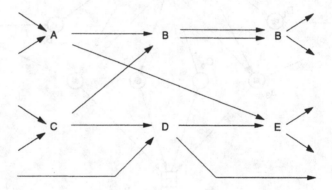

(3)

The upper ingoing and outgoing arcs represent the availability of fork 0, the lower two arcs represent fork 1. "Thinking" is not explicitly represented in (3).

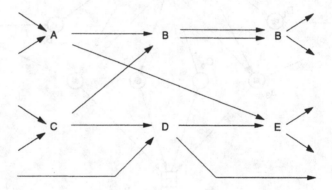

Figure 10.2. A shorthand representation of a concurrent run of $\Sigma_{10.1}$

Eating cycles of other philosophers are likewise abbreviated. In order to represent a concurrent run, those representations are composed in the obvious way. Figure 10.2 thus represents a concurrent run of $\Sigma_{10.1}$. In fact, the interleaved run (1) is one of its interleavings.

In the sequel we distinguish a particularly fair kind of dinner, called *decent* dinners:

> Call a run of $\Sigma_{10.1}$ *decent* iff neighboring philosophers alternate use of their shared fork.   (4)

The runs considered in (1) and in Fig. 10.2 are not decent, because B eats twice consecutively. Hence the fork shared between A and B is not used alternately. This applies correspondingly to the fork shared between B and C.

With the shorthand convention of (2) and (3), Fig. 10.3 shows a concurrent run of $\Sigma_{10.1}$ that is apparently decent. Obviously, a decent concurrent infinite run is uniquely determined by the first use of the forks.

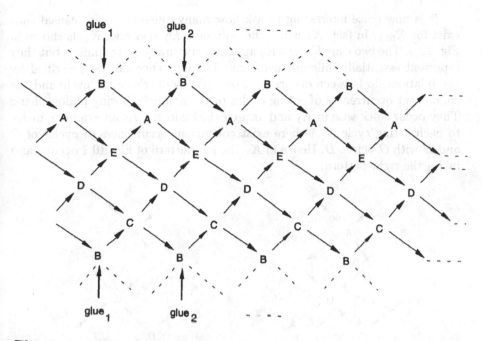

Figure 10.3. Run K_1 of $\Sigma_{10.1}$; employing shorthands as described in (3). Occurrences of B must be identified in the obvious manner.

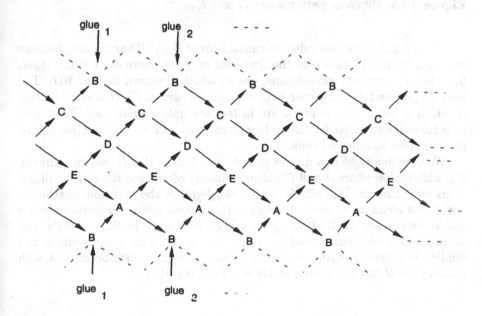

Figure 10.4. Run K_2 of $\Sigma_{10.1}$. Conventions as in Fig. 10.3

It is now quite interesting to ask how many different decent causal runs exist for $\Sigma_{10.1}$. In fact, K_1 is not the only one. Another one, K_2, is shown in Fig. 10.4. The two runs K_1 and K_2 appear structurally quite similar, but they represent essentially different behavior. The difference can be described by the relationship between occurrences of a philosopher's eating cycle and the concurrent occurrences of eating cycles of the non-neighboring philosophers: They occur clock wise in K_1 and anti-clockwise in K_2. As an example, in K_1 to each eating cycle of A there exist concurrent occurrences of cycles of C and D with C before D. Hence in K_1 the left pattern of Fig. 10.1 occurs, and in K_2 the right pattern.

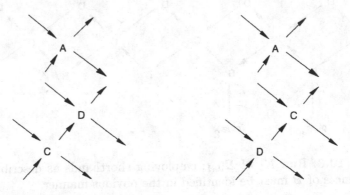

Figure 10.5. Different patterns in K_1 and K_2

Are K_1 and K_2 the only decent causal runs of $\Sigma_{10.1}$? They are not, because due to a "unlucky" choice of the first user of forks, there exist two further, but "less concurrent" causal runs, one of which is shown in Fig. 10.6. The runs K_1, K_2 and K_3, together with the counterpart of K_3 (the construction of which is left to the reader), are in fact the only decent runs. They give structural information on the behavior of $\Sigma_{10.1}$, which can not be gained directly from interleaved runs.

We turn finally to non-decent causal runs. Figure 10.7 shows an example, K_4, with philosophers A and C eating infinitely often, and the other philosophers eating never. The run K_4 sheds new light on the question whether or not B has a chance to grasp his forks. This question is meaningful only if a global view is assumed, allowing for a coincident view at the system's conditions which are represented by a_1 and a_2 in (1). This assumption is not fulfilled in a system with conditions a_1 and a_2 locally distributed and with philosopher B not being able to observe both together.

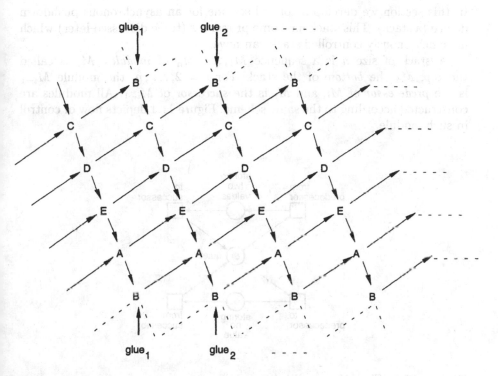

Figure 10.6. Run K_3 of $\Sigma_{10.1}$. Conventions as in Fig. 10.3

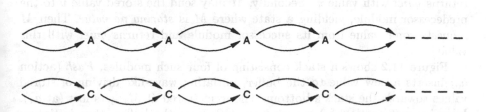

Figure 10.7. Run K_4 of $\Sigma_{10.1}$. Conventions as in Fig. 10.3. Does B get a chance to eat?

11   An Asynchronous Stack

In this section we develop a control scheme for an asynchronous pushdown device (a stack). This stack has some properties (to be discussed later) which no synchronously controlled stack can have.

A stack of size n is a sequence $M_1, \ldots, M_n$ of *modules*. M_1 is called the *top*, M_n the *bottom* of the stack. For $i = 2, \ldots, n$, the module M_{i-1} is the *predecessor* of M_i, and M_i is the *successor* of M_{i-1}. All modules are constructed according to the same scheme. Figure 11.1 depicts flow of control in such module.

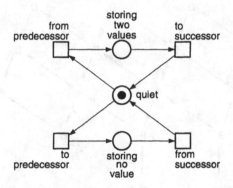

Figure 11.1. Flow of control in a module

In its *quiet* state, a module M stores exactly one value, v. Two alternative actions may occur in this state: Firstly, some value w may arrive from M's predecessor module, yielding a state where M is *storing two values* v and w. Then M propagates the previously held value v to the successor module and returns *quiet* with value w. Secondly, M may send the stored value v to the predecessor module, yielding a state where M is *storing no value*. Then M requests some value from its successor module and returns *quiet* with this value.

Figure 11.2 shows a stack consisting of four such modules. *Push* (action a_0) inserts a new value to the buffer, initiating wave-like driving of stored values towards the stack's bottom. The item stored at M_4 gets lost (at a_4). Likewise, *pop* (action b_0) removes an item from the buffer, thus initiating wave-like popping up of stored values towards the stack's top. M_4 gets some "undefined" value then (by b_4). Each module is assumed to store this "undefined" value initially.

It is intuitively clear that $\Sigma_{11.2}$ in fact models the control structure of a properly behaved stack. It is also obvious how a stack of size n is extended to a

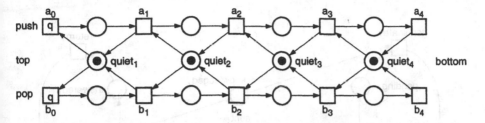

Figure 11.2. Flow of control in the asynchronous stack with capacity for four items

stack of size $n+1$, and that this kind of extension does not affect performance at the top of the stack. Formal arguments specifying those properties and proving them correct will be discussed in Sect. 55.2.

12 Crosstalk Algorithms

In a network of cooperating agents, each agent usually has a distinguished initial state. Each time an agent visits its initial state, it *completes a round*, and its next round is about to begin. A network of agents is said to run a *round policy* (or to be *round-based*) if each message sent in the sender's i-th round is received in the receiver's i-th round. *Crosstalk* arises whenever two agents send each other messages in the same round.

In this section we show what round-based networks of asynchronous, message passing agents may look like. Particular emphasis is given to the issue of crosstalk.

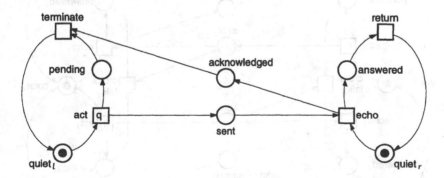

Figure 12.1. Actor and responder

To start with, Fig. 12.1 shows a network of two *sites l* and *r* (the *left* and the *right* site, respectively) and a communication line that links both sites together. In its *quiet* state, l may spontaneously send a message to r and

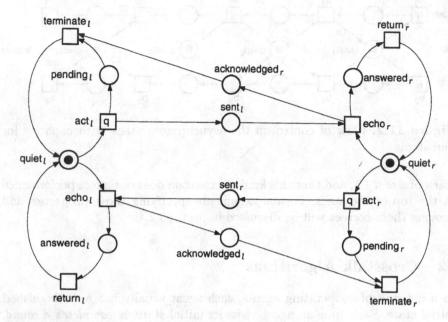

Figure 12.2. Actor/responder sites: deadlock prone

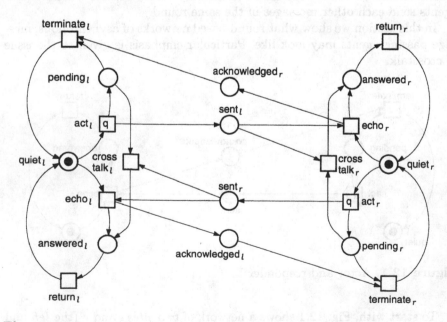

Figure 12.3. Actor/responder sites: round errors possible

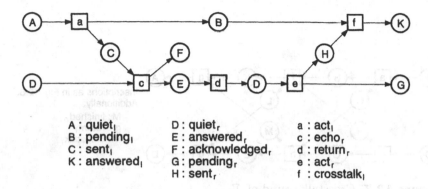

A : quiet$_l$ D : quiet$_r$ a : act$_l$
B : pending$_l$ E : answered$_r$ c : echo$_r$
C : sent$_l$ F : acknowledged$_r$ d : return$_r$
K : answered$_l$ G : pending$_r$ e : act$_r$
                    H : sent$_r$ f : crosstalk$_l$

Figure 12.4. A run of $\Sigma_{12.3}$

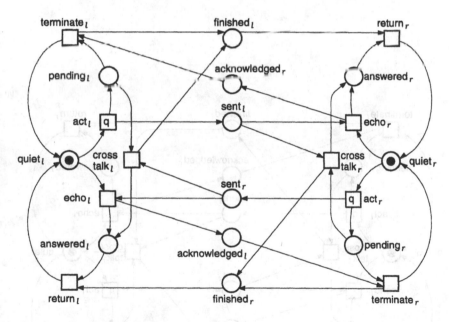

Figure 12.5. Round-based crosstalk

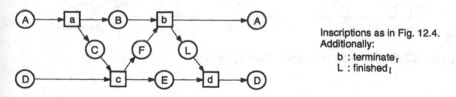

Inscriptions as in Fig. 12.4.
Additionally:
  b : terminate$_r$
  L : finished$_l$

Figure 12.6. Round of $\Sigma_{12.5}$, with actor l and responder r

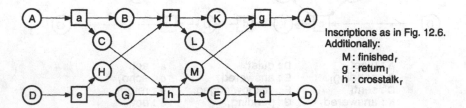

Inscriptions as in Fig. 12.6.
Additionally:
M : finished$_r$
g : return$_l$
h : crosstalk$_r$

Figure 12.7. Crosstalk round of $\Sigma_{12.5}$

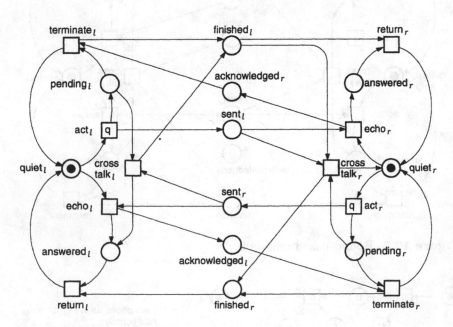

Figure 12.8. Ordered crosstalk: first l, then r

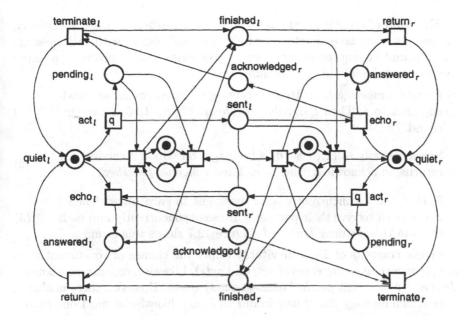

Figure 12.9. Alternately ordered crosstalk

remains in the state *pending* until the receipt of an acknowledgment. Then *l* *terminates* and moves to its *quiet* state, from where *l* may start action again. Upon receiving a message site *r* echoes an acknowledgment, turns *answered* and eventually returns to *quiet*, where *r* is ready to accept the next message. This interplay of the two sites *l* and *r* may be described intuitively in terms of *rounds*. A round starts and ends when both sites are *quiet*. But notice that the site *l* may start its $(i+1)$st round before site *r* has completed its i-th round.

Now we plan to extend the two agents *l* and *r* so as to behave symmetrically (motivated by issues not to be discussed here), i.e., *r* additionally may play the role of a sender and *l* the role of a receiver.

The symmetrical extension $\Sigma_{12.2}$ of $\Sigma_{12.1}$ apparently fails, as the system deadlocks in case both *l* and *r* decide to *act* in the same round: The site *l* in the state *pending$_l$* expects a token on *acknowledged$_r$*, but gets one on *sent$_r$* instead. In this situation, an obvious continuation was to accept the token on *sent$_r$* and to return along *returned$_l$* to *quiet$_l$*. This is achieved by the action *crosstalk$_l$* (and likewise *crosstalk$_r$*) in $\Sigma_{12.3}$. However, $\Sigma_{12.3}$ is not (yet) acceptable: One of its runs, given in Fig. 12.4, is apparently ill-structured: The agent *l* is eventually offered an *acknowledgment* as well as a *message*, and *l* by mistake assumes crosstalk. What actually happened may be called a *round error*: The token on *sent$_r$* belongs to the second round of the system. It reaches *l* before *l* completed the first round, i.e., before *l* properly accepted the first round's acknowledgment.

This kind of error is ruled out in $\Sigma_{12.5}$ by a further message. Intuitively, this message may be understood as a *"round end"* signal, with each message sent in round i being received in round i. Formulated more precisely, a round covers one of the following three sub-runs:

l sends a message to r. Upon receiving it, r returns an acknowledgment to l. Then l signals $finished_l$. Figure 12.6 shows this   (1)
round.

Symmetrically to (1), r sends a message to l. Upon receiving it, l
returns an acknowledgment to r. Then r signals $finished_r$.   (2)

Both l and r concurrently send messages to each other. Then l
and r both receive their partner's message concurrently and each   (3)
of them then returns $finished$. Figure 12.7 shows this round.

The basic concepts of $\Sigma_{12.5}$ inevitably imply the chance of concurrent messages (concurrent occurrences of act_l and act_r). Likewise, concurrent *acknowledged* and concurrent *finished* messages may occur. However, there is always at most one message under way from l to r, and likewise at most one from r to l.

This works perfectly as long as l and r are linked by two physical lines, one for messages from l to r and one for messages from r to l. However, if just *one* line is available, $\Sigma_{12.5}$ may cause mismatch: a message from l to r and a message from r to l may meet on the line. As act_l and act_r are local, quiescent actions, this mismatch can not be ruled out. It can however be *detected* and fixed, provided each sent message, upon meeting some other message at the line, is not entirely destroyed (but only arbitrarily corrupted). For this case, $\Sigma_{12.8}$ *orders* the occurrences of $crosstalk_l$ and $crosstalk_r$: l acts *before* r and consequently $finished_l$ is before $finished_r$. Augmenting tokens on $finished$ with the round's original message then guarantees perfect communication.

For the case of crosstalk this system guarantees $crosstalk_l$ before the corresponding $crosstalk_r$. This is achieved by taking l's round-end message as a further precondition of $crosstalk_r$.

This policy may be considered an unfair preference of l over r. A more symmetrical solution is $\Sigma_{12.9}$, with alternating priority for crosstalk. This system is symmetrical up to an initial bit, with the first crosstalk starting with site l.

13   Mutual Exclusion

Mutual exclusion of local states in a network of cooperating agents is required in a great variety of systems. A lot of phenomena and problems that are typical for distributed systems occur in the attempt to model various concepts and assumptions on mutual exclusion.

This section is intended to glance a couple of those concepts under the aspect of properly modeling mutual exclusion algorithms. Means to formulate and to prove properties of those algorithms will be discussed in Part C.

Two system components (*sites*) are assumed. Each of them includes a particular ("critical") state. The two sites must synchronize such that they always are able to eventually go critical, but never are coincidently in their respective critical state.

The *mutual exclusion problem* is the problem of constructing algorithms achieving the two mentioned requirements. Various assumptions on the sites' capabilities and on the available synchronization mechanisms motivate different solutions.

In the sequel we state the mutual exclusion problem in detail, several solutions will be studied, and their respective advantages and their disadvantages will be discussed.

13.1 The problem

Consider a system essentially consisting of two sites l and r (the *left* and the *right* site). Each site is bound to a cyclic visit of three local states, called *quiet, pending,* and *critical,* as shown in Fig. 13.1, with a quiescent step from *quiet* to *pending* (where the sites' states are indexed l and r, respectively). Two properties are to be guaranteed: firstly, that l' and r never be both together *critical* (the *mutual exclusion* property), and secondly that each *pending* component eventually reaches *critical* and later *quiet* (the *evolution* property).

Figure 13.1. Basic components of *mutex*

A number of well-known algorithms solves this problem, coincidently respecting various additional requirements. For example, it is frequently required that the two sites l and r cooperate in a specific way only; they may share variables or exchange messages or mutually inspect specific local states. Additionally it may be required that a mutual exclusion algorithm is *distributedly implementable.* We refrain from a formal definition of this notion and stick to an apparently necessary condition, the *locality of fairness*: A fair transition together with its pre-set $^\bullet t$ must belong to *one* site and only one

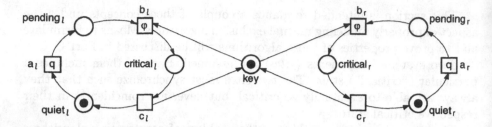

Figure 13.2. The contentious *mutex* algorithm

place $p \in \text{}^\bullet t$ may be forward branching (i.e., $p^\bullet \supsetneq \{t\}$). The partner site may be connected to this place p in a *reading mode* at most, i.e., by loops (p, t') only. This version of fairness is distributedly implementable because conventional hardware guarantees that one site's assignment to a variable is not prevented by the other site's iterated testing of the variable.

First we consider three deficient algorithms, thus pointing out the difficulty of meeting mutual exclusion, evolution, and local fairness at the same time. Then follow four "perfect" algorithms, each with its own merits, and finally we consider two asymmetrical algorithms, granting the left site some kind of preference over the right site.

13.2 The contentious *mutex* algorithm

Figure 13.2 shows an algorithm that in fact meets both requirements of mutual exclusion and evolution. The local state *key*, however, can not be associated uniquely with one of the sites. Both sites compete for *key* and moreover repeated conflict for *key* must be resolved fairly for both partners (as both b_l and b_r are fair actions). Hence, additional global means are necessary to install proper management of *key* nondeterminism. In technical terms, the algorithm neglects locality of fairness for both b_l and b_r and thus is not distributedly implementable.

13.3 The alternating *mutex* algorithm

Figure 13.3 shows an algorithm that respects the requirements of mutual exclusion and local fairness (as no fair transition at all is involved). However, it neglects evolution. For example, the site r may eventually remain *quiet*, in which case the site l may get stuck in its state *pending*. This algorithm may be used in the case of greedy sites only, where both sites strive to go critical as frequently as possible.

13.4 The state testing *mutex* algorithm

Figure 13.4 shows an algorithm with local states *noncrit$_l$* and *noncrit$_r$*, which can be considered as *flags*, allowing the respective partner to go into its critical

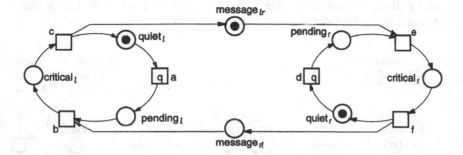

Figure 13.3. The alternating *mutex* algorithm

state. Upon moving to *critical* along b_l, the site l tests the flag $noncrit_r$ and coincidently removes its own flag $noncrit_l$. Occurrence of the action b_l may be prevented forever by infinitely many occurrences of b_r. Hence the assumption of fairness is inevitable for b_l.

The pre-set ${}^\bullet b_l$ of b_l, however, has *two* forward branching elements, $noncrit_l$ and $noncrit_r$, thus violating the requirement of local fairness.

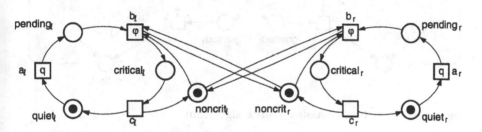

Figure 13.4. The state testing *mutex* algorithm

13.5   The token-passing *mutex* algorithm

Figure 13.5 shows an algorithm based on message passing. The essential concept of the mutex algorithms in Fig. 13.5 is a *token* that at each reachable state is helt by one of the sites. A site may go critical only while holding the token. The site without token may gain it on demand. In $\Sigma_{13.5}$, the token is initially helt by site l in the local state $avail_l$. With the token in $avail_l$, the site l is able to move immediately from $pending_l$ by action a to $critical_l$. With action e the site l then returns from $critical_l$ to $quiet_l$ and makes the token again available for l. Furthermore, l may receive a request for the token sent by the site r along $requested_l$. Fairness of c guarantees that l eventually sends the token to $granted_r$ and coincidently turns $silent_l$. The request sent by site r along $requested_l$ is due to an occurrence of action h. Hence site r is at $waiting_r$ until the token on $granted_r$ arrives.

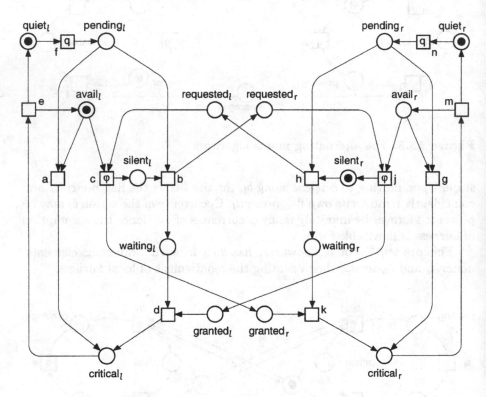

Figure 13.5. A token-passing *mutex* algorithm

Occurrence of action k then brings the site r to $critical_r$. Site l may meanwhile be pending again. As site r is now the owner of the token, site l is in $silent_l$ and may send a request to r by occurrence of b.

The two sites l and r are structurally symmetrical. The initial state, however, is not symmetrical, as site l initially carries the token (at $avail_l$), whereas site r is at $silent_r$. Site l (and likewise site r) is no sequential machine. Actions f and c (actions n and j in site r) may very well occur concurrently.

The two sites cooperate by message passing, with two types of messages ($requested$ and $granted$) in each direction. Site l has one fair action, c. The corresponding conflict place $avail_l$ is not even read by site r. Symmetrically, the conflict places of j is $avail_r$, not read by l.

Site l is fault tolerant with respect to actions a and b. Site r's iterated access to $critical_r$ is not affected in case a or b maliciously remains enabled forever.

13.6 The round-based *mutex* algorithm

The ordered crosstalk algorithm $\Sigma_{12.8}$ can be extended to an algorithm for mutual exclusion, as shown in Fig. 13.6. The ordered occurrence of *crosstalk$_l$* and *crosstalk$_r$* in $\Sigma_{12.8}$ implies that *finished$_l$* and *finished$_r$* never carry a token at the same time. $\Sigma_{13.6}$ refines *finished$_l$* into *crit$_l$*, action n and *terminated$_l$*.

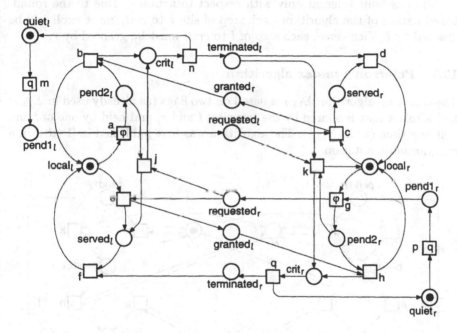

Figure 13.6. A round-based *mutex* algorithm

finished$_r$ is refined correspondingly. Local states and actions of $\Sigma_{10.6}$ are re-named according to their new role in $\Sigma_{13.6}$, and further components (*quiet$_l$*, m, *pend1$_l$*, *quiet$_r$*, p, *pend1$_r$*) are added, providing the elements as required in Fig. 13.1. The system $\Sigma_{13.6}$ operates in rounds: Wanting to go to *critical*, site l sends a request to site r by action a and remains in *pend2$_l$* until site r reacts with a token on either *granted$_r$* or *requested$_r$*. Site l becomes critical in both cases by occurrence of action b or action j, respectively. Site r likewise may send a request to l by action g, then r remains in *pend2$_r$* until site l reacts with a token on *granted$_l$* or *requested$_l$*. Site r becomes critical in the first case by action h. The second case occurs in the situation of crosstalk, where both sites strive at their respective critical state in the same round. Site r has to wait in this case until l leaves *crit$_l$* and sends a token to *terminated$_l$*.

The two sites l and r are structurally not symmetrical: l precedes r in case of crosstalk. Site l is no sequential machine, as n may occur concurrently to e and f. In site r the action q may likewise occur concurrently to c and d.

Fairness of action a guarantees that site l in state $pend_l$ will eventually become critical. The corresponding conflict place, $local_l$, is not read by site r. Symmetrically, the conflict place $local_r$ of the fair action g of site r is not read by site l.

Site l is fault tolerant only with respect to action a. Due to the round-based nature of the algorithm, each step of site r to $crit_r$ must explicitly be granted by l. Vice versa, each step of l to $crit_l$ must be granted by r.

13.7   Peterson's *mutex* algorithm

The following algorithm $\Sigma_{13.7}$ is based on two flags (as already used in $\Sigma_{13.4}$) and a token that is shared by the two sites l and r, and held by one of l and r at any time (as in $\Sigma_{13.5}$). The algorithm was first published in [Pet81] in a programming notation.

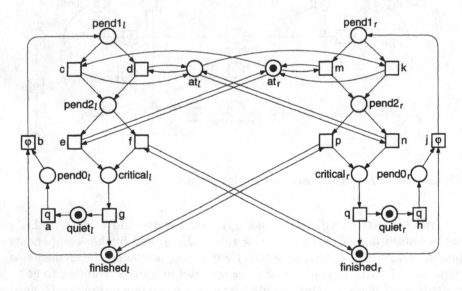

Figure 13.7. Peterson's *mutex* algorithm

The flag $finished_l$ signals to the site r that the site l is presently not striving to become critical. This allows the site r to "easily" access its critical region, by the action p. Likewise, the flag $finished_r$ allows the site l to access its critical state, by the action f. The shared token alternates between at_l and at_r. The step from $pend1_l$ to $pend2_l$ results in the token on at_l: by action c in case l obtains the token from at_r, or by action d in case l held the token anyway.

The step from $pend1_r$ to $pend2_r$ likewise results in the token on at_r. Hence the token is always at the site that executed the step from $pend1$ to $pend2$ most recently.

After leaving $quiet_l$ along the quiescent action a, the site l takes three steps to reach its critical state $critical_l$. In the first step, the fair action b brings l from $pend0_l$ to $pend1_l$ and removes the flag $finished_l$. Fairness of b is local, because ${}^\bullet b = \{pend0_l, finished_l\}$ is local to l, with $finished_l$ the only forward branching place in ${}^\bullet b$, which is connected to the right site, r, by a loop $(finished_l, p)$. The second step, from $pend1_l$ to $pend2_l$, results in the shared token on at_l, as described above. The third step brings l to $critical_l$, with action f in case site r signals with $finished_r$ that it is presently not interested in going critical, or with action e in case the site r more recently executed the step from $pend1_r$ to $pend2_r$. The algorithm's overall structure guarantees that one of $finished_r$ or at_r will eventually carry a token that remains there until eventually either f or e occurs.

The two sites l and r are structurally symmetrical, but the initial state favors the right site.

13.8 Dekker's *mutex* algorithm

The following algorithm $\Sigma_{13.8}$ is a variant of Peterson's algorithm $\Sigma_{13.7}$. It employs the same two flags $finished_l$ and $finished_r$, and likewise shares a token, that is either on at_l or at_r at any time. The essential difference to $\Sigma_{13.7}$ is the time at which the shared token is adjusted: The token is moved to at_l before l becomes critical in $\Sigma_{13.7}$, whereas the token is moved to at_l

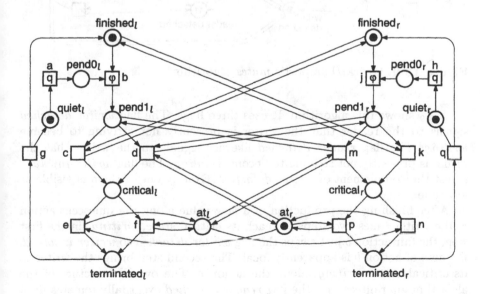

Figure 13.8. Dekker's *mutex* algorithm

after l has been critical in $\Sigma_{13.8}$. In case the site r has not raised its flag *finished$_r$*, the step from *pend1$_l$* to *critical$_l$* with action d depends not only on the shared token in *at$_r$* but also on the local state *pend1$_r$* of site r.

13.9   Owicki/Lamport's *mutex* algorithms

Different sites may be given different priorities, hence different access policies to their respective critical regions. A typical example is a system of a *writer* and a *reader* site of a shared variable: Whenever prepared to update the variable, the writer may eventually execute this update in its critical state. The reader may be guaranteed less: Whenever pending for reading the variable, the reader will eventually get reading or the writer will eventually update the variable. Hence the reader may start to access its critical state in vain.

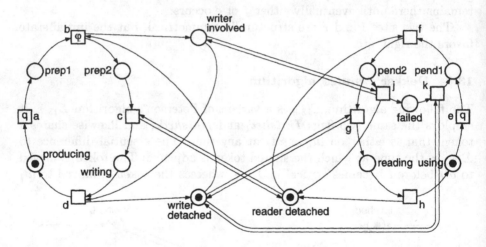

Figure 13.9. Owicki/Lamport's *mutex* algorithm

$\Sigma_{13.9}$ shows this algorithm. It uses three flags: The flag *writer detached* signals to the reader that the writer is presently not striving to become *writing*. The flag *reader detached* likewise signals to the writer that the reader is presently not striving to become *reading*. The flag *writer involved* is just the complement of *writer detached*: Exactly one of them is visible at any time.

After finishing the production of a new value along the quiescent action a, the writer takes two steps to reach its critical state, *writing*. In the first step, the fair action b just swaps the flag *writer detached* to *writer involved*. Fairness of action b is apparently local. The second step brings the writer to its critical state, *writing*, along the action c. The overall structure of the algorithm guarantees that the flag *reader detached* eventually remains until c has occurred.

After using the previous value of the shared variable, the reader may become pending for a new value along the quiescent action e. It takes the reader two steps to reach its critical state, *reading*. Neither of them is guaranteed to occur. Furthermore, the reader in the intermediate state *pend2* may be forced to return to *pend1*. By the first action, f, the reader removes the *reader detached* flag. Iterated occurrence of action c may prevent the occurrence of f (by analogy to the door closing problem of $\Sigma_{6.4}$). The second step, from *pend2* to *reading* with action g, is possible only in case the writer is detached. In case the writer is involved instead, action j releases the flag *reader detached*, allowing the writer to proceed. The reader remains in state *failed* until the writer is detached. In this case, the reader may proceed to *pend1* and start a further attempt to get *reading*.

13.10   The asymmetrical *mutex* algorithm

$\Sigma_{13.10}$ shows a further asymmetrical mutex algorithm that does without any assumption of fairness. Just like the previous algorithm, the prepared writer will eventually proceed to *writing*. The writer, however, may update each newly written value and prevent the reader from reading any value.

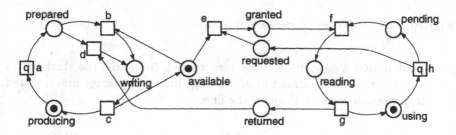

Figure 13.10. The asymmetrical *mutex* algorithm

The algorithm uses three types of messages: *requested* and *returned* sent from the reader to the writer and *granted* sent from the writer to the reader.

After finishing the production of a new value along the quiescent action a, the writer takes either a step via action b or one via action d, to reach its critical state, *writing*. A token on *available* represents the previously written value which not has been read by the reader. Action b or d may yield an updated value. A token on *returned* represents control over the shared variable returned from the reader to the writer, after the reader has read the previous value. Actions a and d then yield a new value.

Along the quiescent action h, the reader, after finishing the use of the previously read value, sends a request for an updated value to the writer. Upon granting a new value along action e, the reader starts reading. However, it may happen that the reader remains stuck in its local state *pending* forever:

The writer either remains producing forever, or the writer produces infinitely many new values and neglects fairness for the action e. The assumption of fairness would help in the latter case.

14   Distributed Testing of Message Lines

Assume a *starter* process s and two *follower* processes, l and r (the *left* and the *right* process, respectively). All three processes are pairwise connected by directed message lines. Figure 14.1 outlines this topology. Each message passing through a line suffers a finite, but unpredictable delay. Processes communicate along those lines only.

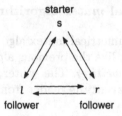

Figure 14.1. Topology of message lines

A distributed algorithm is to be constructed, to enable the starter s to quickly test proper functioning of all message lines. A message line is tested by a test message passing through the line.

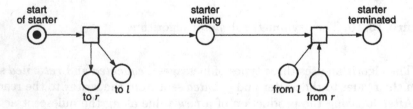

Figure 14.2. Behavior of the starter process s

Figure 14.2 shows the *starter's* behavior: s sends test messages to both l and r, and remains *pending* until receiving test messages from both l and r. Figure 14.3 shows the behavior of the left process l: It starts by receiving a message from s or from r. In the first case, l sends a message to r and remains *waiting* for a message from r. Upon receipt of this message, l sends a message to s and *terminates*. In the second case, after receipt of a message from r, process l sends a message to s and remains *waiting* for a message

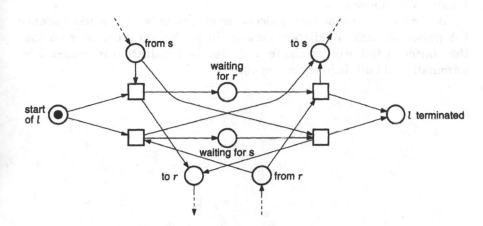

Figure 14.3. Behavior of follower process *l*

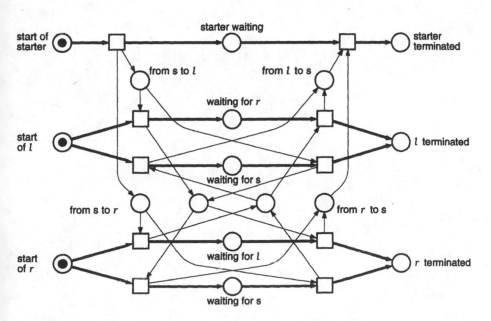

Figure 14.4. Test algorithm for network (1) (with boldfaced arcs for agents *starter*, *left*, and *right*)

from s. Upon receipt of this message, l sends a message to r and *terminates*. Finally, s terminates, too.

It is easy to see that each process terminates only after a test message has passed through all adjacent message lines. We will prove later on that the starter in fact will terminate and that its termination is preceded by termination of both follower processes.

Part B
Advanced System Models

Part A introduced a formalism coping with the essentials of concurrency. Its expressive power will be increased in this part, allowing the integration of data structures and the concise representation of unhandily large elementary net systems. A technique for modeling real, large systems results. Two aspects will govern this procedure: Firstly, new concepts are introduced as *specializations (refinements)* of existing ones. Hence, all notions already introduced translate canonically to the new case. Secondly, powerful *analysis techniques* should be available for the new formalism. Such techniques will be presented in Part D.

III. Advanced Concepts

This chapter provides the central basis of the modeling technique of this book: the concept of *system nets*.

The step from elementary to general system nets can be understood in two different ways. Firstly, as a *generalization*: While elementary system nets stick to (distributed) control structure, general system nets additionally provide data structures. Technically, the dynamic elements (tokens) in a net are no longer black dots, but any kind of data.

The second view of general system nets conceives them as *shorthand* or *concise representations* of elementary system nets: Multiple occurrences of similarly structured subnets are *folded* to a single net structure. Its various instances (unfoldings) are characterized by *inscriptions* of the net elements. This approach is particularly suitable, because all notions and concepts of es-nets translate canonically to system nets. It goes without saying that it is in general not intended to unfold a system net explicitly. Any kind of reasoning on system nets will be executed without explicit unfolding.

15   Introductory Examples

Three motivating examples will be presented in this section. Technical details follow in Sect. 16.

15.1   The producer/consumer system revisited

We return to the very first net model of a producer/consumer system, as displayed in Figs. 1.1 and 8.1. This net describes production, delivery, removal, and consumption of *any* item. No concrete, specific item has been named. Now we assume a specific item, *a*; Figure 15.1 represents the producer/consumer system for the object *a*. In the state shown, the action *produce a* is enabled, and its occurrence yields the state shown in Fig. 15.2. Due to the inscription "*a*" at the arc linking *produce a* and *ready to deliver a*, the token to occur at *ready to deliver a* is no longer a black dot, but the item *a*. The action *deliver a* is enabled in this state, because the two ingoing arcs start from local states that carry items according to the arcs' inscriptions: *buffer empty* carries a

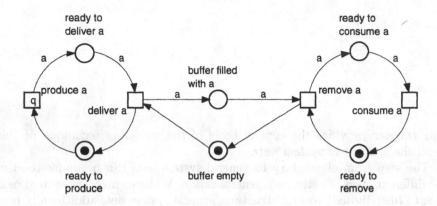

Figure 15.1. Producing and consuming objects of sort *a*

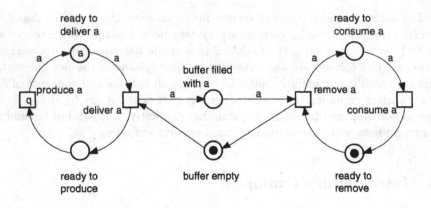

Figure 15.2. After occurrence of *produce a*

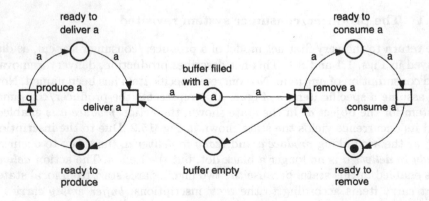

Figure 15.3. After occurrence of *deliver a*

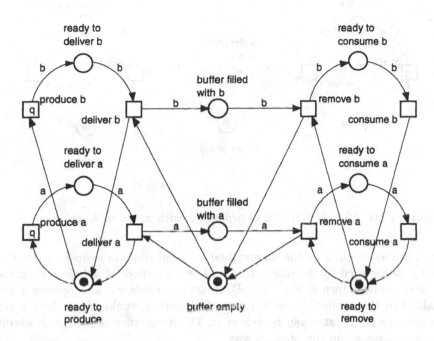

Figure 15.4. Producing and consuming objects a or b

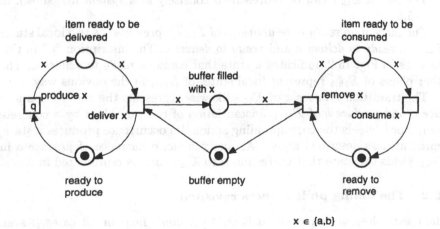

Figure 15.5. Producing and consuming any kind of items

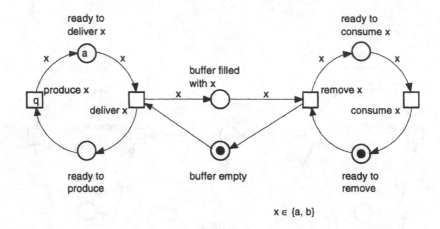

Figure 15.6. After occurrence of *produce* x with $x = a$ in $\Sigma_{15.5}$

black dot as required by the uninscribed arc, and *ready to deliver* carries the item a as required by the inscription a. The occurrence of *deliver* a then reveals the state shown in Fig. 15.3. Both actions *produce* a and *remove* a are enabled in this state. The occurrence of *produce* a would cause the second appearance of "a" at *ready to deliver* a. The occurrence of *remove* a would enable *consume* a in the obvious way.

Figure 15.4 now extends $\Sigma_{15.1}$ to enable the treatment of a second item, b. In the state shown, there is a choice between *produce* a and *produce* b. Choice of *produce* a would cause the behavior described above. Choice of *produce* b would correspondingly cause processing of the item b.

The es-net $\Sigma_{15.4}$ can be represented concisely as a *system net*, shown in Fig. 15.5.

The place *item ready to be delivered* of $\Sigma_{15.5}$ represents the two local states of $\Sigma_{15.4}$, *ready to deliver* a and *ready to deliver* b. The inscription "a" of this place, as in Fig. 15.2, indicates a state that contains *ready to deliver* a. The other places of $\Sigma_{15.5}$ represent local states of $\Sigma_{15.4}$ in the obvious way.

The transition *produce* x of $\Sigma_{15.5}$ likewise represents the two actions *produce* a and *produce* b of $\Sigma_{15.4}$. Instantiation of the variable x by a concrete item, a or b, yields the corresponding action. Its occurrence produces a state, represented as described above. As an example, occurrence of *produce* a in $\Sigma_{15.5}$ yields the state that corresponds to $\Sigma_{15.2}$, and is represented in $\Sigma_{15.6}$.

15.2   The dining philosophers revisited

Plain variables, as employed in Sect. 15.1, don't help in all cases. As an example, we return to the system of thinking and eating philosophers, as introduced in Sect. 10. For the sake of simplicity we stick to the case of *three* philosophers, and for reasons to become obvious soon, we redraw the corresponding es-net, as shown in Fig. 15.7.

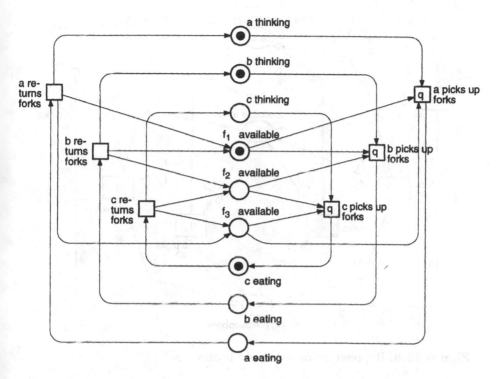

Figure 15.7. The system of three philosophers

We strive at a more concise representation of the system by exploiting its regular structure. The essential idea is to represent a set s of local states with "similar" behavior as a single place p and likewise a set of actions with "similar" behavior as a single transition.

As an example, the three local states "a is thinking", "b is thinking", and "c is thinking" in $\Sigma_{15.7}$ may be assigned the place "thinking philosophers". A state with a and b thinking and c not thinking then corresponds to a state where "thinking philosophers" is inscribed with a and b, but not with c. a and b are then in the *actual extension* of the place "thinking philosophers", whereas c is not.

In the framework of es-nets, the local states may change upon the occurrence of actions. This corresponds to a change in the extension of the corresponding place.

Figure 15.8 shows a corresponding net: The local states of $\Sigma_{15.7}$ are now clustered into three places, "thinking philosophers", "available forks", and "eating philosophers", respectively. The extension of each place is given by its inscription. Each action is now to indicate the items affected by its occurrence. This is achieved by inscriptions of the corresponding arcs. As an example, the inscriptions of the arcs adjacent to "a picks up" in $\Sigma_{15.8}$ indicate that upon the occurrence of this transition, the forks f_1 and f_3 leave "available

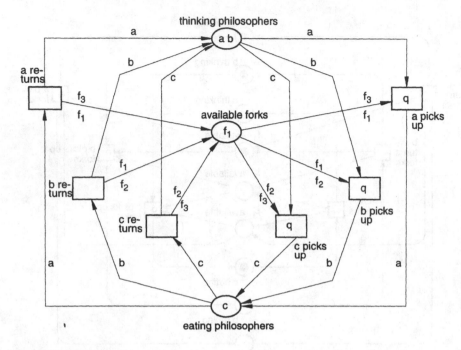

Figure 15.8. Representation using predicates

forks", and likewise the philosopher a leaves "thinking philosophers" and enters "eating philosophers".

In a further transformation, and for reasons to become obvious shortly, we replace the inscriptions of $\Sigma_{15.8}$ as shown in Fig. 15.9: The two forks employed by any philosopher x are denoted by $l(x)$ and $r(x)$. Hence, l and r represent functions, assigning each philosopher x his left and right fork, respectively.

In a final step, a set of actions is *folded* to a single transition. As an example, the three actions "a picks up", "b picks up", and "c picks up" of $\Sigma_{15.9}$ are represented by the transition "pick up" in Fig. 15.10. Return of forks is represented correspondingly.

The instance of a distinguished action (e.g., "a picks up") is in the folded version represented by an assignment of concrete items to the variables occurring at the surrounding arcs. As an example, "a picks up" corresponds in $\Sigma_{15.10}$ to the assignment of a to the variable x. In fact, this assignment yields the inscriptions of arcs surrounding "a picks up" in $\Sigma_{15.9}$.

With $\Sigma_{15.10}$ we have obtained a more abstract and general representation.

15.3   The distributed sieve of Eratosthenes

Here we consider the well-known example of identifying the prime numbers in a set of integers, according to the *Sieve of Eratosthenes*. The conventional

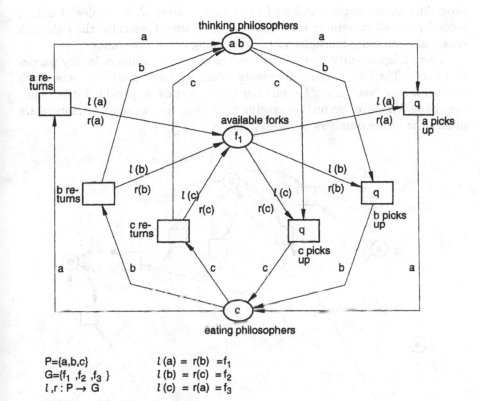

P={a,b,c} $l(a) = r(b) = f_1$
G={f_1, f_2, f_3} $l(b) = r(c) = f_2$
$l,r : P \rightarrow G$ $l(c) = r(a) = f_3$

Figure 15.9. Representation using functions

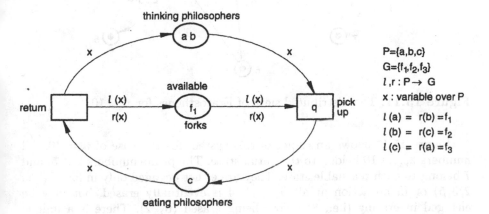

P={a,b,c}
G={f_1,f_2,f_3}
$l,r : P \rightarrow G$
x : variable over P

$l(a) = r(b) = f_1$
$l(b) = r(c) = f_2$
$l(c) = r(a) = f_3$

Figure 15.10. Representation using parameterized actions

procedure traverses the numbers from 2 to n, erasing all multiples of 2. In a second path, all remaining multiples of 3 are erased. Generally, the i-th path erases all remaining multiples of the remaining $(i + 1)$-st number.

There is apparently no need to erase multiples of numbers in any particular order. The following requirements suffice for erasing all products: Each number k may "see" any other number and may erase it, provided it is a multiple of k. Whenever no further erasing is feasible, the remaining numbers are in fact the prime numbers between 2 and n.

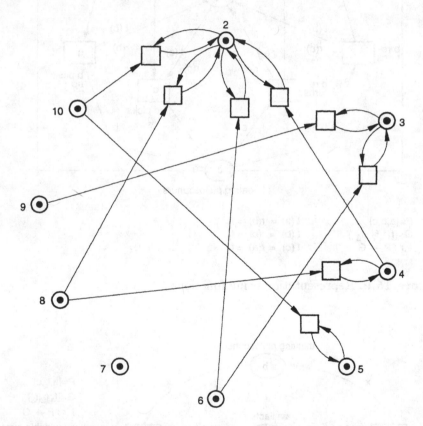

Figure 15.11. The distributed sieve of Eratosthenes for $n = 10$

Figure 15.11 shows an es-net of this system for the case of $n = 10$: All numbers $2, \ldots, 10$ belong to the initial state. The prime numbers $2, 3, 5$, and 7 belong to each reachable state: They are either engaged only in loops (i.e., $2, 3, 5$) or in no action at all (i.e., 7). 4 is eventually erased, but may be engaged in erasing (i.e., 8) before being erased (by 2). There is a unique number, 3, to erase 9. Numbers 6, 8, and 10 may be erased alternatively by two numbers, respectively.

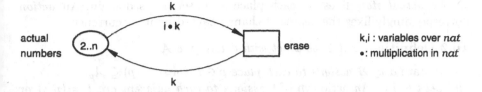

Figure 15.12. The distributed sieve of Eratosthenes for any number n

Figure 15.12 shows the folded version for any number, n: *actual numbers* initially carries all numbers from 2 to n. Transition *erase* is enabled whenever some number k and some multiple $i \cdot k$ of k (for $i > 1$) is present in *actual numbers*. Both k and $i \cdot k$ are removed, and k returns to *actual numbers*.

15.4 Conclusion

The above introductory examples followed a simple idea to concisely represent elementary system nets: Firstly, the local states are partitioned into several classes. Each class is then "folded" to a single place. Such a place contains particular *items*, representing the local states in the corresponding class. Likewise, the actions are partitioned into classes, and each class is folded to a single transition. Inscriptions of the adjacent arcs describe the actions in the corresponding class. A distinguished action can be regained by evaluating the variables involved (i.e., by replacing them with constants). Nets of this kind are examples of *system nets*.

A formal framework for system nets has to establish the relationship between syntactical inscriptions (terms), at arcs and places, and their concrete semantical denotation. This relationship of syntax and semantics is mathematically well established, belonging to the basic concepts of computer science. Furthermore, it is intuitively obvious, as the above examples show. At first reading one may therefore proceed to Sect. 17 immediately.

16 The Concept of System Nets

The conceptual idea of system nets is quite simple: Each place of a system net Σ represents a *set* of local states and each transition of Σ represents a *set* of actions. The sets assigned to the places form the underlying universe:

16.1 Definition. *Let Σ be a net. A universe $\mathcal{A}$ of Σ fixes for each place $p \in P_\Sigma$ a set $\mathcal{A}_p$, the domain of p in $\mathcal{A}$.*

For example, in Fig. 15.10 the domain of *thinking philosophers* and of *eating philosophers* is the set of philosophers, and the domain of *available forks* is the set of forks.

An *actual state* fixes for each place a subset of its domain. An *action* correspondingly fixes the degree of change caused by its occurrence:

16.2 Definition. *Let Σ be a net with a universe $\mathcal{A}$.*

i. *A state a of Σ assigns to each place $p \in P_\Sigma$ a set $a(p) \subseteq \mathcal{A}_p$.*
ii. *Let $t \in T_\Sigma$. An action m of t assigns to each adjacent arc $f = (p,t)$ or $f = (t,p)$ a set $m(f) \subseteq \mathcal{A}_p$.*

In fact, the state of $\Sigma_{15.10}$ as shown in Fig. 15.10 assigns to *thinking philosophers* the set $\{a, b\}$, to *available forks* the set $\{f_1\}$ and to *eating philosophers* the set $\{c\}$. A typical action m of *return* in $\Sigma_{15.10}$ was given by $m(\textit{eating philosophers}, \textit{return}) = m(\textit{return}, \textit{thinking philosophers}) = \{c\}$ and $m(\textit{return}, \textit{available forks}) = \{f_2, f_3\}$.

Concession and effect of actions, and the notion of steps, are defined in correspondence to Def. 3.1:

16.3 Definition. *Let Σ be a net with some universe $\mathcal{A}$, let a be a state, let $t \in T_\Sigma$, and let m be an action of t.*

i. *m has concession (is enabled) at a iff for each place $p \in {}^\bullet t$, $m(p,t) \subseteq a(p)$ and for each place $p \in t^\bullet$, $(m(t,p) \setminus m(p,t)) \subseteq \mathcal{A}_p \setminus a(p)$.*
ii. *The state $\mathrm{eff}(a,m)$, defined for each place $p \in P_\Sigma$ by*

$$
\mathrm{eff}(a,m)(p) := \begin{cases} a(p) \setminus m(p,t) & \textit{iff } p \in {}^\bullet t \setminus t^\bullet, \\ a(p) \cup m(t,p) & \textit{iff } p \in t^\bullet \setminus {}^\bullet t, \\ (a(p) \setminus m(p,t)) \cup m(t,p) & \textit{iff } p \in t^\bullet \cap {}^\bullet t, \\ a(p) & \textit{otherwise,} \end{cases}
$$

is the effect of m's occurrence on a.
iii. *Assume m is enabled at a. Then the triple $(a, m, \mathrm{eff}(a,m))$ is called a step of t in Σ, and usually written $a \xrightarrow{m} \mathrm{eff}(a,m)$.*

The action m described above of the transition *return* is enabled at the state shown in Fig. 15.10 (m is moreover the only enabled action). Its occurrence yields the state with all philosophers thinking, all forks available, and no philosopher eating.

Steps may be described concisely by means of a canonical extension of actions:

16.4 Proposition. *Let Σ be an es-net, let $t \in T_\Sigma$, and let $a \xrightarrow{m} b$ be a step of t. For all $(r,s) \notin F_\Sigma$, extend m by $m(r,s) := \emptyset$. Then for all places $p \in P_\Sigma$, $b(p) = (a(p) \cup m(t,p)) \setminus m(p,t)$.*

System nets are now defined by analogy to elementary system nets in Sect. 8: A net with a domain for each place and a set of actions for each transition is furthermore equipped with an initial state, and distinguished subsets of quiescent and fair transitions.

16.5 Definition. *A net Σ is a* system net *iff*

  i. For each place $p \in P_\Sigma$, a set $\mathcal{A}_p$ is assumed (i.e., a universe of Σ),
  ii. for each transition $t \in T_\Sigma$, a set of actions of t *is assumed,*
  iii. a state a_Σ is distinguished, called the initial state *of Σ,*
  iv. each transition $t \in T_\Sigma$ is denoted as either progressing *or* quiescent,
  v. some progressing transitions are distinguished as fair.

The introductory examples in Sect. 15 employ a particular representation technique for system nets: The initial state is given by place inscriptions, and the actions of a transition are given by the valuations of the variables as they occur in arc inscriptions. Details will follow in Sect. 19.

17 Interleaved and Concurrent Runs

Interleaved runs of system nets can be defined canonically as sequences of steps. There is likewise a canonical definition of concurrent runs, corresponding to Def. 5.4.

Based on the notion of steps given in Def. 16.3, interleaved runs are defined by analogy to Def. 4.1:

17.1 Definition. *Let Σ be a system net.*

  i. For $i = 1, \ldots, n$ assume steps $a_{i-1} \xrightarrow{m_i} a_i$ of Σ. They form a Σ-based finite interleaved run w, *written $a_0 \xrightarrow{m_1} a_1 \xrightarrow{m_2} \ldots \xrightarrow{m_n} a_n$. Each $i \in \{0, \ldots, n\}$ is an* index *of w.*
  ii. For $i = 1, 2, \ldots$ assume steps $a_{i-1} \xrightarrow{m_i} a_i$ of Σ. They form a Σ-based infinite interleaved run w, *sometimes outlined $a_0 \xrightarrow{m_1} a_1 \xrightarrow{m_2} \ldots$. Each $i \in N$ is an* index *of w.*

For example, Fig. 17.1 shows an interleaved run of $\Sigma_{15.12}$ for the case of $n = 10$: Each state a is represented by listing $a(A)$ in a column. Each action m is represented as a pair (k, l), with $m(A, t) = \{k, l \cdot k\}$ and $m(t, A) = \{k\}$.

2	2	2	2	2	2
3	3	3	3	3	3
4 $\xrightarrow{(5,2)}$	4 $\xrightarrow{(3,3)}$	4 $\xrightarrow{(4,2)}$	4 $\xrightarrow{(3,2)}$	4 $\xrightarrow{(2,2)}$	5
5	5	5	5	5	7
6	6	6	6	7	
7	7	7	7		
8	8	8			
9	9				
10					

Figure 17.1. Interleaved run of $\Sigma_{15.12}$

Reachable steps, states and actions are defined in analogy to Def. 8.3:

17.2 Definition. *Let Σ be a system net.*

i. *A step $a \xrightarrow{m} b$ of Σ is reachable in Σ iff there exists a finite interleaved run $a_\Sigma \xrightarrow{m_1} a_1 \xrightarrow{m_2} a_2 \to \ldots \to a_{n-1} \xrightarrow{m_n} a_n$ with $a_{n-1} \xrightarrow{m_n} a_n = a \xrightarrow{m} b$.*
ii. *A state a of Σ is reachable in Σ iff $a = a_\Sigma$ or there exists a reachable step formed $b \xrightarrow{m} a$.*
iii. *An action m is reachable in Σ iff there exists a reachable step formed $a \xrightarrow{m} b$.*

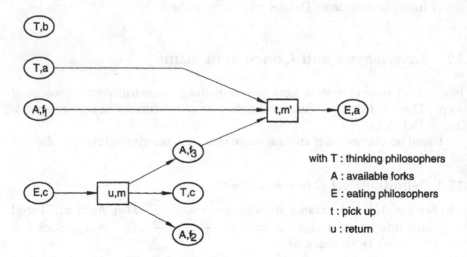

Figure 17.2. Concurrent run of $\Sigma_{15.10}$

Concurrent runs are now defined in two stages: Firstly, each action m is assigned an *action net*, representing the action's details in terms of an inscribed net. In a second step, those nets are "glued together", forming a concurrent run.

17.3 Definition. *Let Σ be a system net, let $t \in T_\Sigma$, let m be an action of t, and let N be an injectively labeled net with $T_N = \{e\}$. Furthermore, assume $l(e) = (t, m)$, $l(^\bullet e) = \{(p, a) \mid p \in {}^\bullet t,$ and $a \in m(p, t)\}$, $l(e^\bullet) = \{(p, a) \mid p \in t^\bullet,$ and $a \in m(t, p)\}$. Then N is an action net of Σ (for m).*

For example,

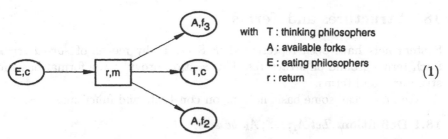

with T : thinking philosophers
A : available forks
E : eating philosophers   (1)
r : return

is an action net for the action m of $\Sigma_{15.10}$ with $m(E, r) = \{c\}$, $m(r, A) = \{f_2, f_3\}$ and $m(r, T) = \{c\}$.

17.4 Definition. *Let Σ be a system net and let K be an element labeled occurrence net. K is a Σ-based concurrent run iff*

   i. in each concurrent state a of K, different elements of a are differently labeled,

  ii. for each $t \in T_K$, $({}^{\bullet}t \cup t^{\bullet}, \{t\}, {}^{\bullet}t \times \{t\} \cup \{t\} \times t^{\bullet})$ is an action net of Σ.

As an example, Fig. 17.2 shows a concurrent run of $\Sigma_{15.10}$. The involved actions m and m' are obvious from the context. As a further example, Fig. 17.3 shows a concurrent run of $\Sigma_{15.12}$ for the case of $n = 10$. Each place label (A, i) is depicted by i and each transition label (t, m) is represented as a pair (k, l) with $m(A, t) = \{k, k \cdot l\}$ and $m(t, A) = \{k\}$.

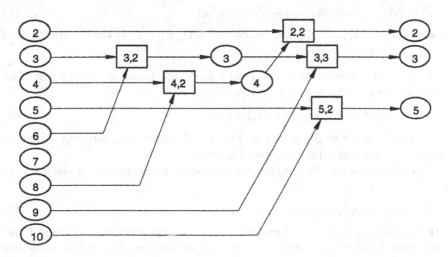

Figure 17.3. Concurrent run of $\Sigma_{15.12}$ (place inscriptions (A, i) and transition inscriptions $(t, (k, i))$ are represented by i and k, i, respectively.)

18   Structures and Terms

System nets have been represented in Sect. 15 by means of *sorted terms*. Such terms ground on *structures*. This section provides the formal basis for structures and terms.

We first recall some basic notions on constants and functions:

18.1 Definition. *Let $A_1, \ldots, A_k$ be sets.*

  i. Let $a \in A_i$ for some $1 \le i \le k$. Then a is called a constant in the sets *$A_1, \ldots, A_k$ and A_i is called a* sort *of a.*

  ii. For $i = 1, \ldots, n+1$ let $B_i \in \{A_1, \ldots, A_k\}$, and let $f : B_1 \times \ldots \times B_n \to B_{n+1}$ be a function. Then f is called a function over the sets *$A_1, \ldots, A_k$. The sets $B_1, \ldots, B_n$ are the* argument sorts *and B_{n+1} is the* target sort *of f. The $n+1$-tuple $(B_1, \ldots, B_{n+1})$ is the* arity *of f and is usually written $B_1 \times \ldots \times B_n \to B_{n+1}$.*

For example, in Fig. 15.10, b is a constant in P and G of sort P. Furthermore, l is a function over P and G with one argument sort P and the target sort G. Its arity is $P \to G$.

A *structure* is just a collection of constants and functions over some sets:

18.2 Definition. *Let $A_1, \ldots, A_k$ be sets, let $a_1, \ldots, a_l$ be constants in $A_1, \ldots, A_k$ and let $f_1, \ldots, f_m$ be functions over $A_1, \ldots, A_k$. Then*

$$A = (A_1, \ldots, A_k; a_1, \ldots, a_l; f_1, \ldots, f_m) \tag{1}$$

is a structure. *$A_1, \ldots, A_k$ are the* carrier sets, *$a_1, \ldots, a_l$ the* constants, *and $f_1, \ldots, f_m$ the* functions *of A.*

In fact, the system nets $\Sigma_{15.4}$ and $\Sigma_{15.6}$ are based on structures. The structure for the philosophers system $\Sigma_{15.10}$ is

$$Phils = (P, G; a, b, c; f_1, f_2, f_3; l, r) \tag{2}$$

with P, G, l, and r as described in Fig. 15.10. Hence this structure has two carrier sets, six constants, and two functions.

The structure for the concurrent version of Eratosthenes' n-sieve $\Sigma_{15.12}$ is

$$Primes = (\mathbb{N}; 2, \ldots, n; \cdot) \tag{3}$$

with $\mathbb{N}$ denoting the natural numbers, $2, \ldots, n$ the numbers between 2 and n for some fixed $n \in \mathbb{N}$, and $\cdot$ the product of integers. Hence this structure has one carrier set, $n - 1$ constants, and one function.

The composition of functions of a structure can be described intuitively by means of *terms*. To this end, each constant a of a structure A is represented by a *constant symbol* a and likewise each function f of A by a *function symbol* f. (This choice of symbols is just a matter of convenience and convention. Any other choice of symbols would do the same job). Furthermore, terms include *variables*:

18.3 Definition. *Let* $\mathcal{A} = (A_1, \ldots, A_k; a_1, \ldots, a_l; f_1, \ldots, f_m)$ *be a structure.*

  i. *Let* $X_1, \ldots, X_k$ *be pairwise disjoint sets of symbols. For* $x \in X_i$, *call* A_i *the* sort *of* x $(i = 1, \ldots, k)$. *Then* $X = X_1 \cup \ldots \cup X_k$ *is a set of* $\mathcal{A}$-sorted *variables.*

  ii. *Let* X *be a set of* $\mathcal{A}$-sorted variables. For all $B \in \{A_1, \ldots, A_k\}$ *we define the sets* $T_B(X)$ *of terms of sort* B *over* X *inductively as follows:*
    a) $X_i \subseteq T_{A_i}$;
    b) *for all* $1 \leq i \leq l$, *if* B *is the sort of* a_i *then* $\mathbf{a_i} \in T_B(X)$.
    c) *For all* $1 \leq i \leq m$, *if* $B_1 \times \ldots \times B_n \to B$ *is the arity of* f_i *and if* $t_j \in T_{B_j}(X)$ $(j = 1, \ldots, n)$ *then* $\mathbf{f}(t_1, \ldots, t_n) \in T_B(X)$.

  iii. *The set* $T_{\mathcal{A}}(X) := T_{A_1}(X) \cup \ldots \cup T_{A_k}(X)$ *is called the set of* $\mathcal{A}$-terms *over* X.

For example, with respect to the two structures *Phils* and *Primes* considered above, 1(b) and r(x) are Phils-terms of sort G over $\{x\}$, where the sort of x is P. Likewise, $2 \cdot 5$ and $3 \cdot y$ are *Primes*-terms of sort N over $\{y\}$, where the sort of y is N.

In the sequel we always assume some (arbitrarily chosen, but) fixed *order* on variables. Generally we use the following notation:

18.4 Notation. *A set* M *is said to be* ordered *if a unique tuple* $(m_1, \ldots, m_k)$ *of pairwise different elements* m_i *is assumed such that* $M = \{m_1, \ldots, m_k\}$. *We write* $M = (m_1, \ldots, m_k)$ *in this case.*

Each term u over an ordered set of sorted variables describes a unique function, *val*^{u}, the *valuation* of u:

18.5 Definition. *Let* $\mathcal{A}$ *be a structure and let* $X = (x_1, \ldots, x_n)$ *be an ordered set of* $\mathcal{A}$-sorted variables. For $i = 1, \ldots, n$ *let* B_i *be the sort of* x_i *and let* $u \in T_B(X)$ *for any sort* B *of* $\mathcal{A}$. *Then* $B_1 \times \ldots \times B_n$ *is the set of arguments for* X *and the valuation of* u *in* $\mathcal{A}$ *is a function* $val^u : B_1 \times \ldots \times B_n \to B$, *which is inductively defined over the structure of* u:

$$val^u(a_1, \ldots, a_n) = \begin{cases} a_i & \text{if } u = x_i \text{ for } 1 \leq i \leq n, \\ a & \text{if } u = \mathbf{a} \text{ for some constant } a \text{ of } \mathcal{A}, \\ f(val^{u_1}(a_1, \ldots, a_n), \ldots, val^{u_k}(a_1, \ldots, a_n)) \\ \quad \text{if } u = \mathbf{f}(u_1, \ldots, u_k) \text{ for some function} \\ \quad f \text{ of } \mathcal{A} \text{ and terms } u_1, \ldots, u_k \in T_{\mathcal{A}}(X). \end{cases}$$

For example, with respect to the structure *Primes* considered above, $u = (2 \cdot y) \cdot x$ is a *Primes*-term over $X = \{x, y\}$. Assuming X is ordered $X = (x, y)$, we get $val^u(3, 4) = val^{2 \cdot y}(3, 4) \cdot val^a(3, 4) = val^2(3, 4) \cdot val^y(3, 4) \cdot 3 = (2 \cdot 4) \cdot 3 = 8 \cdot 3 = 24$. As a special case we consider terms without variables:

18.6 Definition. *Let* $\mathcal{A}$ *be a structure.*

i. *The set* $T_A(\emptyset)$ *consists of the* A-*ground terms and is usually written* T_A.

ii. *For each* $u \in T_A$ *of sort* B, val^u *is the unique function* $val^u : \emptyset \to B$, *i.e.*, val^u *indicates a unique element in* B. *This element will be denoted* val^u.

For example, with respect to the structure *Phils* considered above, $u = 1(b)$ is a *phils*-ground term with $val^u = val^{f_3} = f_3$.

This completes the collection of notions and notations to deal with structures and terms.

19   A Term Representation of System Nets

Based on structures and terms as introduced in the previous section, a representation of system nets is suggested in the sequel, as already used in Sect. 15. The representation of a transition's actions is the essential concept. To this end, each transition t is assigned its set M_t of *occurrence modes*. Each occurrence mode then defines an action. A typical example was

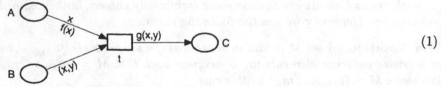

$$(1)$$

Assume the variable x is of sort M, y of sort N and x ordered before y. Then $M \times N$ is the set of occurrence modes of t. Each pair $(m,n) \in M \times N$ defines an action $\widetilde{mn}$ of t, gained by substituting m and n for x and y in the adjacent terms. Hence $\widetilde{mn}(A,t) = \{m, f(m)\}$, $\widetilde{mn}(B,t) = \{(m,n)\}$ and $\widetilde{mn}(t,C) = \{g(m,n)\}$.

The syntactical representation of term-based system nets reads as follows:

19.1 Definition. *Let* Σ *be a net and let* A *be a structure. Assume*

i. *each place* $p \in P_\Sigma$ *is assigned a carrier set* A_p *of* A *and a set* $a_\Sigma(p) \subseteq T_{A_p}$ *of ground terms,*

ii. *each transition* $t \in T_\Sigma$ *is assigned an ordered set* X_t *of* A-*sorted variables,*

iii. *each arc* $f = (t,p)$ *or* $f = (p,t)$ *adjacent to a transition* t *is assigned a set* $\overline{f} \subseteq T_{A_p}(X_t)$ *of* A_p-*terms over* X_t;

iv. *each transition* $t \in T_\Sigma$ *is denoted either* progressing *or* quiescent, *and some progressing transitions are distinguished as* fair.

Then Σ *is called a* term inscribed *over* A.

In graphical representations, the places p and the arcs (r,s) are inscribed by $a_\Sigma(p)$ and $\overline{rs}$, respectively. Figures 15.1–15.5, 15.9, and 15.11 show examples. Occurrence modes and actions of a transition are defined as follows:

19.2 Definition. *Let Σ be a term inscribed net and let $t \in T_\Sigma$ be a transition.*

  i. *Let $(x_1, \ldots, x_n)$ be the ordered set of variables of t and let M_i be the sort of x_i $(i = 1, \ldots, n)$. Then $M_t := M_1 \times \ldots \times M_n$ is the set of occurrence modes of t.*
 ii. *Let $m \in M_t$. For each adjacent arc $f = (p, t)$ or $f = (t, p)$ and different $u, v \in \bar{f}$ assume $val^u(m) \neq val^v(m)$. Then $\tilde{m}$ is an action of t, defined by $\tilde{m}(f) = \{val^u(m) \mid u \in \bar{f}\}$.*

The action $\widetilde{mn}$ discussed above is in fact an action of the transition (1). A term-inscribed net obviously represents a system net:

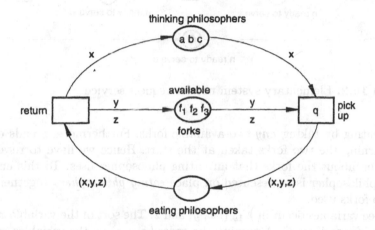

Figure 19.1. A variant of $\Sigma_{15.10}$

19.3 Definition. *Let Σ be a net that is term-inscribed over a structure $\mathcal{A}$ such that for all $p \in P_\Sigma$ and all different $u, v \in a_\Sigma(p)$, $val^u \neq val^v$. Then the system net of Σ consists of*

– *the universe $\mathcal{A}$,*
– *for all $t \in T_\Sigma$, the actions of t as defined in Def. 19.2(ii),*
– *the initial state a, defined for each place $p \in P_\Sigma$ by*
   $a(p) := \{val^u \mid u \in a_\Sigma(p)\},$
– *the quiescent, progressing, and fair transitions, as defined by Σ.*

As a variant of the term-represented philosophers system $\Sigma_{15.9}$ we consider a more liberal access policy to the available forks in $\Sigma_{19.1}$: Assume that the available forks lie in the middle of the table. Each philosopher p

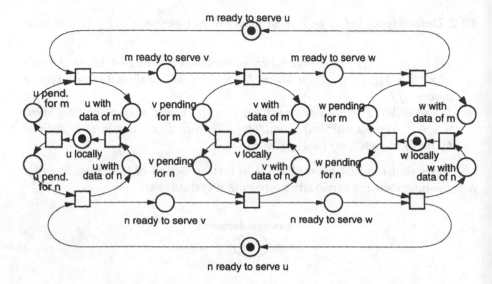

Figure 19.2. Elementary system net for request service

starts eating by taking *any* two available forks. Furthermore, *p* ends eating by returning the two forks taken at the start. Hence we have to retain information about the forks that an eating philosopher uses. To this end, an eating philosopher is represented on place *eating philosophers* together with the two forks used.

Three variables occur in $\Sigma_{19.1}$: x, y, and z. The sort of the variable x is P, the sort of y and z is G. Assuming the order (x, y, z) on the variables, we get $M_{return} = M_{pick\ up} = P \times G \times G$. With the occurrence mode $m = (b, f_1, f_3)$, the action $\widetilde{m}$ of *pick up* is given by $\widetilde{m}(thinking\ philosophers, pick\ up) = \{b\}$, $\widetilde{m}(available\ forks, pick\ up) = \{f_1, f_3\}$, and $\widetilde{m}(pick\ up, eating\ philosophers)$ $= \{(b, f_1, f_3)\}$. This action is enabled at $a_{\Sigma_{19.1}}$. Its occurrence then yields the step $a_{\Sigma_{19.1}} \xrightarrow{\widetilde{m}} s$ with $s(thinking\ philosophers) = \{a, c\}$, $s(available\ forks) = \{f_2\}$, and $s(eating\ philosophers) = \{(b, f_1, f_3)\}$.

As a further example we consider a simple algorithm for deterministic distributed request service, as shown in the elementary system net of Fig. 19.2. Three data users u, v, and w are to be served by two data managers m and n in cyclic order. Initially, each data user works locally. After some time he requires data from both data managers. Upon being served by both m and n, the data user returns to local work. Each data manager in a cycle first serves u, followed by v and w.

Figure 19.3 gives a system net representation of this system. The underlying structure is

$$(Users, Managers, Users \times Managers, Managers \times Users, u, v, w, m, n, suc) \quad (1)$$

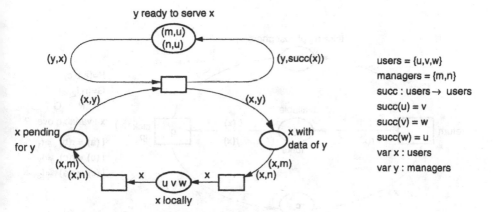

Figure 19.3. A distributed request service

with details given in Fig. 19.3. In this example, arc inscriptions such as (x, m) are terms including variables (e.g., x) as well as constant symbols (e.g., m).

20 Set-Valued Terms

The formalism of Sect. 19 is adequate for many system nets. But there exist more general system nets requiring *set-valued* terms. In order to specify this issue more precisely, assume an es-net Σ with a transition $t \in T_\Sigma$, an action m of t, and a place $p \in {}^\bullet t \cup t^\bullet$ with domain A. Then $\widetilde{m}(p, t)$ or $\widetilde{m}(t, p)$ is a subset of A, with each single term $u \in \overline{pt}$ or $u \in \overline{tp}$ contributing a single element, $val^u(m) \in A$. Now we suggest single terms v that contribute a *subset* $val^v(m) \subseteq A$. More precisely, *set-valued constant symbols, set-valued function symbols*, and *set-valued variables* will be used. We start with motivating examples for all three types of terms.

As a first example we return to the representation of the philosophers system in Fig. 15.10. The graphical representation there of three philosophers a, b, c and three forks f_1, f_2, f_3 is reasonable and lucid. The corresponding system with say, 10, philosophers and 10 forks would become graphically monstrous and for 100 or more items this kind of representation is certainly no longer adequate.

It is better to employ *set-valued constant symbols* P and G. The valuation of P returns the set $P = \{a, b, c\}$ of philosophers and the valuation of G returns the set $G = \{f_1, f_2, f_3\}$ of forks. Figure 20.1 thus shows a typical application of those symbols P and G.

The next example motivates the use of *set-valued function symbols*: All versions of the philosophers system considered so far assigns *exactly two* forks to each eating philosopher. Now we follow the policy as represented in $\Sigma_{20.2}$: Philosopher a eats with one fork, f_1, and philosopher b with two forks, f_2 and

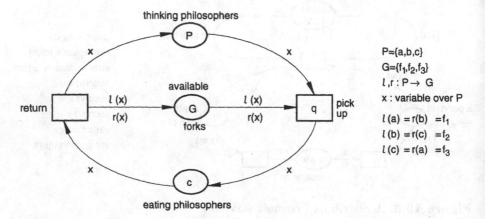

Figure 20.1. Set-valued constant symbols P and G

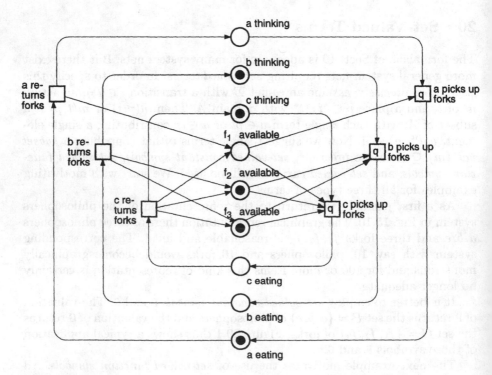

Figure 20.2. Three philosophers with different numbers of forks

f_3. Philosopher c, finally, employs all three forks. A concise representation of this behavior must be based on the function $\Phi : P \to \mathcal{P}(G)$ with $\Phi(a) = \{g_1\}$, $\Phi(b) = \{g_2, g_3\}$, and $\Phi(c) = \{g_1, g_2, g_3\}$. This function can not be described by a set of functions $f : P \to G$. So we employ a *set-valued* function symbol Φ of arity $P \to \mathcal{P}(G)$, with $val^\Phi(p) = \Phi(p)$ for each $p \in P$. $\Sigma_{20.3}$ employs this function symbol. A typical run of this system is given in Fig. 20.4.

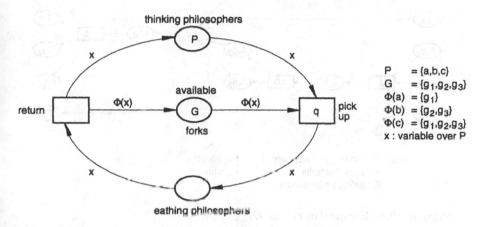

Figure 20.3. System net corresponding to $\Sigma_{20.2}$

The last item to be motivated is the use of set-valued variables. As an example we combine $\Sigma_{19.1}$ and $\Sigma_{20.3}$ into the most liberal access policy of philosophers to forks: A philosopher chooses *any* set of forks each time he starts eating. This case is frequently denoted as the *drinking philosophers* system: The philosophers drink cocktails in a bar. The bar essentially consists of a stock of bottles. When he wants a cocktail, a philosopher takes some of the bottles from the stock, takes them to his place, mixes a cocktail, drinks it, and then returns the bottles. The same philosopher may choose a different bottles for each cocktail. Figure 20.4 represents this behavior, using a set-valued variable, Y, of sort *set of bottles*.

The most general case includes both kinds of terms: Element-valued terms such as a, b, c, x, y, z in $\Sigma_{19.1}$ and $l(x)$, $r(x)$ in $\Sigma_{20.1}$, and set-valued terms such as P, G, $\Phi(x)$ in $\Sigma_{20.3}$, and Y, (x, Y) in $\Sigma_{20.5}$. For the sake of uniform management of both cases, the evaluation $val^u(m)$ of terms u will be slightly adjusted, yielding a set $setval^u(m)$ in any case:

20.1 Definition. *Let Σ be a term inscribed net over a structure $\mathcal{A}$.*

i. Let $p \in P_\Sigma$ and let $u \in a_\Sigma(p)$. Then

$$setval^u = \begin{cases} \{val^u\} & \textit{if the sort of } u \textit{ is } A_p \\ val^u & \textit{if the sort of } u \textit{ is } \mathcal{P}(A_p). \end{cases}$$

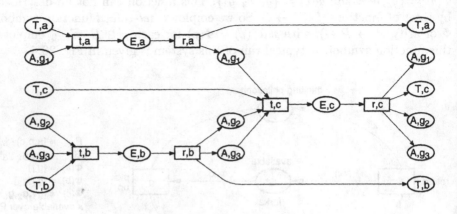

with T : thinking philosophers   t : pick up
      A : available forks   r : return
      E : eating philosophers

Figure 20.4. Concurrent run of $\Sigma_{20.3}$

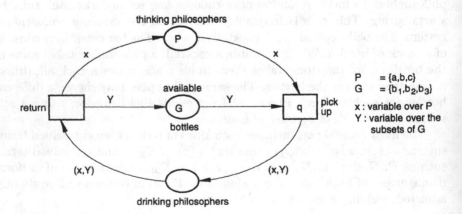

Figure 20.5. Drinking philosophers

ii. Let $f = (p,t) \in F_\Sigma$ or $f = (t,p) \in F_\Sigma$, let $u \in \bar{f}$, and let m be an argument of X_t. Then

$$\text{setval}^u(m) = \begin{cases} \{val^u(m)\} & \text{if the sort of } u \text{ is } A_p \\ val^u(m) & \text{if the sort of } u \text{ is } \mathcal{P}(A_p). \end{cases}$$

For example, in $\Sigma_{20.3}$ we obtain $\text{setval}^x(a) = \{a\}$ for $a \in P$ and $\text{setval}^{\Phi(x)}(b) = \{g_1, g_2\}$.

The actions of a term inscribed net with both element-valued and set-valued terms is now defined as follows:

20.2 Definition. Let Σ be a term inscribed net, let $t \in T_\Sigma$, and let $m \in M_t$. For each adjacent arc $f = (p,t)$ or $f = (t,p)$ and different $u, v \in \bar{f}$ assume $\text{setval}^u(m) \cap \text{setval}^v(m) = \emptyset$. Then $\tilde{m}$ is an action of t, defined by $\tilde{m}(f) = \bigcup_{u \in \bar{f}} \text{setval}^u(m)$.

As an example, the one-element ordered set $\{x\}$ is the set of variables of *pick up* in $\Sigma_{20.3}$. Hence b is an occurrence mode of *pick up*. The action $\tilde{b}$ is then defined by $\tilde{b}(\text{thinking philosophers}) = \tilde{b}(\text{eating philosophers}) = \{b\}$ and $\tilde{b}(\text{available forks}) = \{f_2, f_3\}$. Likewise, let $(x, 1')$ be the set of variables of t in $\Sigma_{20.5}$. Then $m = (b, \{f_1, f_3\})$ is an occurrence mode of *pick up*. Then $\tilde{m}$ is defined by $\tilde{m}(\text{thinking philosophers}) = \{b\}$, $\tilde{m}(\text{available bottles}) = \{f_1, f_3\}$ and $\tilde{m}(\text{drinking philosophers}) = (b, \{f_1, f_3\})$. A further occurrence mode of $\Sigma_{20.5}$ was, e.g., $(b, \{f_2\})$.

20.3 Proposition. Let Σ be a term inscribed net, let $t \in T_\Sigma$, let m be an action of t, and let a be a state of Σ. For all $(r,s) \notin F_\Sigma$ let $\overline{rs} := \emptyset$.

   i. m is enabled at a iff, for each $p \in P_\Sigma$, $\bigcup_{u \in \overline{pt}} \text{setval}^u(m) \subseteq a(p)$ and $(\bigcup_{u \in \overline{tp}} \text{setval}^u(m) \setminus \bigcup_{u \in \overline{pt}} \text{setval}^u(m)) \cap a(p) = \emptyset$.

   ii. Let $a \xrightarrow{m} b$ be a step of Σ. Then for each $p \in P_\Sigma$, $b(p) = (a(p) \setminus \bigcup_{u \in \overline{pt}} \text{setval}^u(m)) \cup \bigcup_{u \in \overline{tp}} \text{setval}^u(m)$.

Proof. i. $\{val^u(m) \mid u \in \overline{tp} \setminus \overline{pt}\} = \bigcup_{u \in \overline{tp}} \text{setval}^u(m) \setminus \bigcup_{u \in \overline{pt}} \text{setval}^u(m)$ by Def. 20.1(ii).

ii. By Def. 20.1(ii) and Proposition 16.4.   □

The system net of a term-inscribed net with both element-valued and set-valued terms is defined as a conservative extension of the corresponding notion in Sect. 19.3 for element-valued terms:

20.4 Definition. Let Σ be a net that is term-inscribed over a structure $\mathcal{A}$, such that for all $p \in P_\Sigma$ and all different $u, v \in a_\Sigma(p)$ holds $\text{setval}^u \cap \text{setval}^v = \emptyset$. Then the system net of Σ consists of

- the universe of $\mathcal{A}$,
- for all $t \in T_\Sigma$, the actions of t as defined in Def. 20.2,

– the initial state a, defined for each place $p \in P_\Sigma$ by $a(p) := \bigcup_{u \in a_\Sigma(p)} \text{setval}^u$,
– the quiescent, progressing and fair transitions, as defined by Σ.

Figures 20.2 and 20.3 in fact show system nets.

21 Transition Guards and System Schemata

21.1 Transition guards

As in sequential programs, a decision between alternative actions frequently depends on data. For example, if an integer x and a data item y are produced independently, then processing y may continue in either of two ways, depending on whether x is positive or negative.

An intuitively conventional representation for this structure was

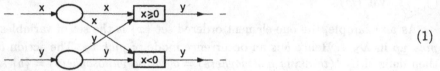

$$(1)$$

But so far system nets do not include transition inscriptions such as "$x \geq 0$" or "$x < 0$". However, such inscriptions can easily be augmented as shorthands to avoid loops. The classical representation for (1) then was

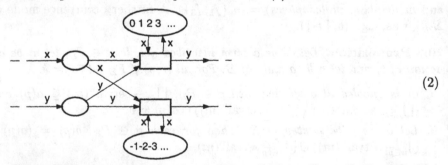

$$(2)$$

Generally, each transition t of a term-inscribed net Σ may be inscribed by a term u that involves variables of t only. Each occurrence mode m then must yield a truth value $val^u(m) \in \{\text{true}, \text{false}\}$: An interleaved or concurrent run then must consist only of actions $\tilde{m}$ with $val^u(m) = \text{true}$.

This concept is so obvious that we refrain from a formal definition.

Transition guards are quite useful, as in the following example of the *distributed predicate meeting problem*.

This slightly abstract problem assumes a function $f : \mathbb{N} \to \mathbb{N}$ and a predicate $Q \subseteq \mathbb{N}$. The task is to find any $i \in \mathbb{N}$ such that $f(i) \in Q$. ("Q holds at $f(i)$").

A sequential solution is shown in $\Sigma_{21.1}$, testing $Q(f(1)), Q(f(2)), \ldots$. This solution is turned to a distributed solution in $\Sigma_{21.2}$ for any $n \in \mathbb{N}$. Intuitively, n concurrent "strands of computation" try to find some proper $i \in \mathbb{N}$.

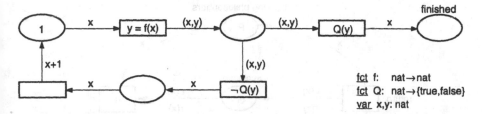

Figure 21.1. Sequential solution to the predicate holding problem

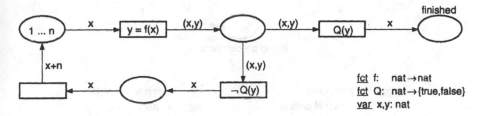

Figure 21.2. Distributed solution to the predicate holding problem

This solution raises the problem of "stopping" all strands after the success of one strand. In a truly concurrent setting this is an amazingly involved question.

21.2 System schemata

Different term-inscribed Petri nets may operate with the same terms. As an example, a variant of $\Sigma_{20.1}$ may operate with

$$P = \{a, b\} \qquad\qquad l(a) = r(b) = f_1$$
$$G = \{f_1, f_2, f_3, f_4\} \qquad r(a) = f_2 \qquad\qquad\qquad (3)$$
$$l, r : P \to G \qquad\qquad l(b) = f_3$$

The net with all its inscriptions may remain in this case. There are just the involved symbols P, G, l, and r that are evaluated in different ways. The inscribed net is just a *schema* for any system that works with two sorts of items, called *philosophers* and *forks*, two constant sets P and G of philosophers and forks, respectively, and two functions that assign a fork to each philosopher. This system schema is represented in Fig. 21.3.

Generally, a system schema is a term-inscribed net with the underlying structure not entirely fixed. Thus, a system schema represents a *set* of system nets. A representation of a system schema declares some *sorts* (domains) and some constants, functions, and variables over standard sorts, declared sorts, cartesian products, or powersets of sorts. We furthermore assume standard sorts such as the natural numbers *nat* or the truth values *bool*, together with the usual operations. Some additional requirements may focus the intended interpretations.

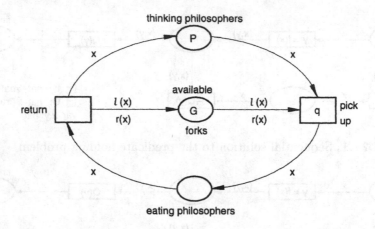

Figure 21.3. A system schema

Most distributed algorithms, as considered in the forthcoming chapters, will be represented as system schemata.

IV. Case Studies

A broad choice of distributed algorithms, modeled as system nets, will be discussed in this chapter, by analogy to the algorithms modeled as elementary system nets in Chap. II.

We start out with some extensions of elementary system models from Chap. II, followed by the paradigm of *constraint programming* which is particularly amenable to distributed execution. Then follow algorithms that organize distributed database updates, consensus, communication over unreliable channels, and anonymous networks.

22 High-Level Extensions of Elementary Net Models

Many of the algorithms considered in Chap. II can naturally be generalized, mainly to aspects of data handling, and thus are adequately modeled as system nets. For the simplest producer/consumer system this has been carried out already for the motivating examples of Sect. 15. Here we start with slightly more involved producer/consumer systems.

22.1 Producer/consumer systems

The introductory example $\Sigma_{1.1}$ was extended in Figs. 9.1 and 9.2 from one to two buffer cells, and in $\Sigma_{15.5}$ from control to data aspects. Here we extend $\Sigma_{15.5}$ furthermore to the case of n buffer cells, with n any natural number. In analogy to $\Sigma_{9.1}$ and $\Sigma_{9.2}$, buffers can be organized sequentially or concurrently.

In order to describe (by analogy to Sect. 9.1) a system with a sequential buffer of some length n, the system schema of Fig. 22.1 employs pairs (a, i) representing item a to be stored at the i-th buffer cell. Action *forward* forwards items from cell i to cell $i + 1$. Any reasonable interpretation of this system schema should evaluate the constant n as a natural number and assign natural numbers to the variable i. The addition symbol $+$ is to be interpreted as addition in N.

It might be worth remarking that in concurrent runs of this system, the action *forward* may concurrently occur in different modes (e.g., mode $x = a$, $i = 1$ and $x = b$, $i = 2$).

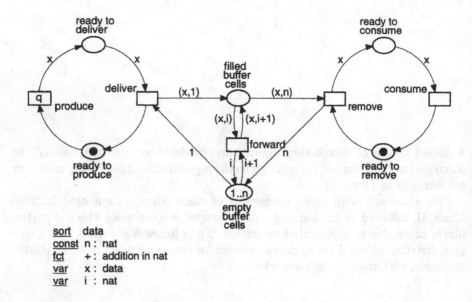

Figure 22.1. A sequential buffer with n cells

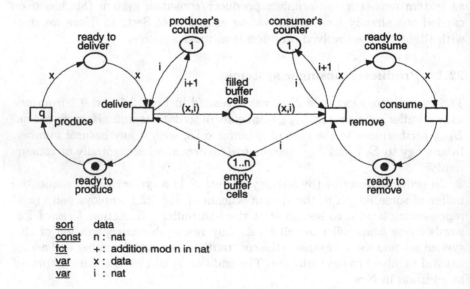

Figure 22.2. A parallel buffer with n cells

A parallel version of a buffer by analogy to Fig. 9.3 is shown in Fig. 22.2: The producer employs a counter for selecting the next buffer cell. If this cell is empty (corresponding token in *empty buffer cells*), an item may go to this cell by occurrence of *deliver*. The consumer likewise employs a counter to select the next buffer cell for removal of an item.

22.2 Decent philosophers

We are now behind an extension of $\Sigma_{15.10}$ that implements decent behavior by philosophers. To this end the place *priority* represents each fork by the pair of its potential users, with the next user mentioned first. For example, (A, B) denotes the fork shared by A and B, with A to use the fork next. (B, A) denotes the same fork, but with B as its next user.

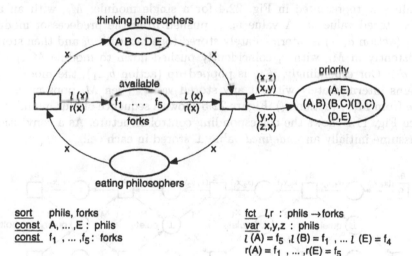

sort	phils, forks	fct	l, r : phils $\rightarrow$ forks
const	A, ... ,E : phils	var	x,y,z : phils
const	f_1 , ... ,f_5: forks		$l(A) = f_5$,$l(B) = f_1$, ... $l(E) = f_4$
			$r(A) = f_1$, ... ,$r(E) = f_5$

Figure 22.3. Decent philosophers

Now $\Sigma_{21.3}$ is given a new initial state and extended by the place *priority* to represent each fork's next user, as in Fig. 22.3. In the state shown there, A and D are the philosophers to start eating, and in fact the only behavior possible in $\Sigma_{22.3}$ is shown in Fig. 10.3.

22.3 Asynchronous stack

A control schema for an asynchronous stack was suggested in Sect. 11. This will now be extended to cover also data flow. By analogy to Fig. 11.1 flow

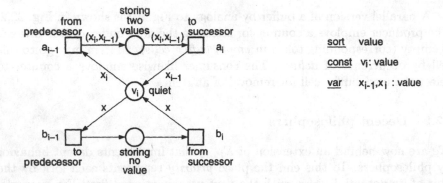

Figure 22.4. Module M_i

of values is represented in Fig. 22.4 for a single module, M_i, with an actually stored value, v_i. A value v_{i-1}, pushed from the predecessor module M_{i-1} (action a_{i-1}) is intermediately stored together with v_i and then stored persistently in M_i, with v_i coincidently pushed down to module M_{i+1} (action a_i). Correspondingly, if v_i is popped up (action b_{i-1}), the module M_i remains intermediately without any stored value. Then M_i pops up a new value from M_{i+1} (action a_i). Figure 22.5 shows a sequence of four modules, of which Fig. 11.2 shows the corresponding control structure. As a convention, we assume initially an *undefined* value $\perp$ stored in each cell.

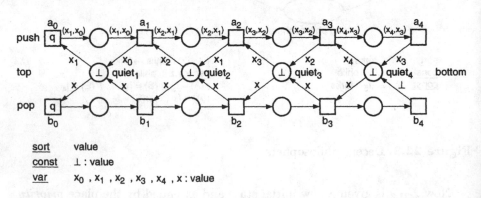

Figure 22.5. Asynchronous stack with capacity for four items

The regular structure of $\Sigma_{22.5}$ allows a parameterized representation of the "inner" modules M_2 and M_3 and furthermore a generalization to n modules, as in Fig. 22.6.

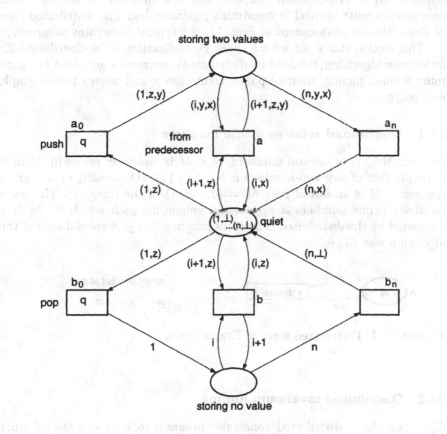

sort   value   var   x, y, z : value
const   ⊥: value   var   i : nat
const   n : nat

Figure 22.6. Asynchronous stack with capacity for n items

23   Distributed Constraint Programming

The paradigm of constraint programming advocates the concept of starting out with a broad domain of candidates for a problem's solution. This domain then is constrained during a program's execution. Elements of the domain may be extinguished independently of each other. Hence concurrent execution is quite natural in constraint programming. The distributed sieve of Eratosthenes, as discussed in Sect. 15, is a typical constraint program.

This section starts out with a slight generalization of the distributed Eratosthenes algorithm, followed by distributed constraint algorithms for maximum finding, sorting, shortest paths, connectivity, and convex hull of graphs and polygons.

23.1   Distributed relative prime numbers

In a set $M \subseteq N$ of natural numbers, $m \in M$ is *relatively prime* in M iff m is no product of any two numbers $a, b \in M \setminus \{1\}$. (Obviously, m is a prime number if M is an initial part $M = \{1, \ldots, n\}$ of the integers). The set of relatively prime numbers is apparently unique for each set $M \subseteq N$. It is computed by the distributed constraint program $\Sigma_{23.1}$. A special case of this algorithm was $\Sigma_{15.12}$.

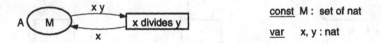

const M : set of nat

var   x, y : nat

Figure 23.1. Distributed sieve of Eratosthenes

23.2   Distributed maximum finding

$\Sigma_{23.2}$ provides a distributed constraint program to compute the maximal element of any finite set of numbers. The function symbol *max* denotes the function that returns the greater of two argument numbers. Each occurrence of transition t considers two elements and eliminates the smaller one.

The algorithm terminates with one element at place A, which then is the maximal element.

const M : set of nat
fct   max : nat × nat → nat
var   x, y : nat

Figure 23.2. Distributed maximum finding

23.3   Distributed sorting

Assume a set of n indexed cards, each holding a *content*, i.e., an alphabetic string and its actual *index*, i.e., a natural number. Initially each number $1, \ldots, n$ is the index of exactly one card. The task is to re-arrange the indices such that eventually the alphabetic order of content coincides with the numerical order of indices.

$\Sigma_{23.3}$ provides a distributed solution to this problem: Essentially, the indices of two cards are swapped in case the context order disagrees with the index order. The algorithm terminates with the index cards holding sorted indices.

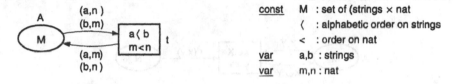

Figure 23.3. Distributed sorting

23.4   Distributed shortest path

Assume a finite, directed graph with each arc labeled by some non-negative number, called its *weight*. Let $u \xrightarrow{a} v$ denote an arc from node u to node v with weight a. Arcs $v_{i-1} \xrightarrow{a_i} v_i (i = 1, \ldots k)$ form a *path from v_0 to v_k with weight* $a_1 + \ldots + a_k$. For any two nodes u and v, the *distance from u to v* is infinite in case there exists no path from u to v. Otherwise it is the smallest of all weights of paths from u to v. The task is to compute for each pair (u, v) of nodes the distance from u to v.

$\Sigma_{23.4}$ shows a distributed solution to this problem. M denotes the initially given graph, representing each arc $u \xrightarrow{a} v$ as (u, v, a). If no arc from u to v exists, M contains the triple (u, v, ∞). Transition t replaces the actual distance c of an entry (u, w, c) by a smaller distance $a + b$ in case a corresponding path from u along some node v to node w was found. The algorithm terminates with triples (u, v, a) at p, giving the distance a to the pair (u, v) of nodes.

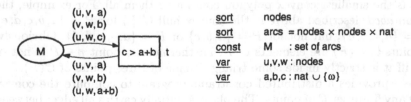

Figure 23.4. Distributed shortest path

23.5 Distributed connectivity

A finite undirected graph is said to be *connected* if each two nodes u and v are linked (along several other nodes) by a sequence of arcs.

$\Sigma_{23.5}$ provides a distributed constraint program to decide whether or not a given graph is connected. The constant M consists of the singleton sets $\{u_1\}, \ldots, \{u_k\}$ of the graph's nodes $u_1, \ldots, u_k$. Each arc linking two nodes u and v is in the net represented as (u, v) or as (v, u).

Transition t constructs sets of nodes that represent connected subgraphs. If t can no longer be enabled, the contents of A decide the problem: The graph G is connected if and only if A finally contains *one* set (which then consists of all nodes of G).

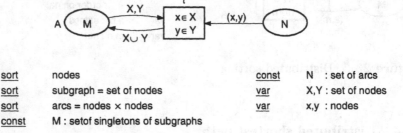

sort	nodes	const	N : set of arcs
sort	subgraph = set of nodes	var	X,Y : set of nodes
sort	arcs = nodes × nodes	var	x,y : nodes
const	M : setof singletons of subgraphs		

Figure 23.5. Distributed connectivity

23.6 Distributed convex hull

A *polygon* in the plane is a finite sequence $a_0 \ldots a_n$ of points in the plane. A polygon defines an *area* with edges $(a_0, a_1), (a_1, a_2), \ldots, (a_{n-1}, a_n), (a_n, a_0)$. As an example, Fig. 23.6 outlines the area of the polygon *abcd*. The points e and f are situated inside and outside this area, respectively. A polygon is *convex* if each edge linking any two points is entirely inside the polygon's area. For example, in Fig. 23.6 the outlined polygon *abcd* is convex, whereas *aebc* is not.

Each finite set of P of points in the plane is assigned its *convex hull* $C(P)$, which is the smallest convex polygon containing them all. For example, the polygon *abcd* described above is the convex hull $C(P)$ for $P = \{a, b, c, d, e\}$ or $P = \{a, b, c, d\}$, but not for $P = \{a, b, c\}$ or $P = \{a, b, c, d, f\}$. Obviously, the points of $C(P)$ are elements of P. Furthermore, a point $p \in P$ is not in $C(P)$ iff it is strictly inside some triangle made of three points of $C(P)$.

$\Sigma_{23.7}$ provides a distributed constraint program to compute the convex hull of any finite set P of points. The place A initially carries all edges between nodes of P. The predicate $inside(a, b, c, d)$ returns *true* iff the point d is inside the triangle abc. Hence transition t eliminates all edges connected to some point inside a triangle. The algorithm terminates with the edges of the convex hull of P at place A.

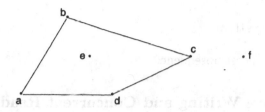

Figure 23.6. A polygon

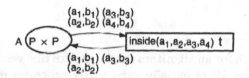

sort	point	fct	inside : point × point × point × point → bool
const	P : set of points	var	$a_1,a_2,a_3,a_4,b_1,b_2,b_3,b_4$: point

Figure 23.7. Distributed convex hull

23.7 Longest upsequence problem

Let $\sigma = a_1 \ldots a_n$ be a sequence of n numbers. Let $1 \leq i_1 < i_2 < \ldots < i_k = n$ be an increasing sequence of k indices of σ. Then $(a_{i_1}, \ldots, a_{i_k})$ is an *upsequence* of σ iff $a_{i_1} < a_{i_2} < \ldots < a_{i_k}$ (notice that $a_{i_k} = a_n$). Clearly there exists a (not necessarily unique) longest upsequence of σ. Let $up(\sigma)$ denote its length.

The *upsequence length problem* is the problem of computing $up(a_1 \ldots a_n)$ for each index $n \in N$ of any (infinite) sequence $a_1, a_2, \ldots$ of numbers.

$\Sigma_{23.8}$ solves this problem. Its essential component is place q holding triples (n, x, j). Each such triple states that the value of a_n equals x and that $up(a_1 \ldots a_n) \geq j$. Action t generates those triples by nondeterministically choosing the value x, picking the next index n from place p and initializing j by 1. "Better" values are computed by u: if the actual value of $up(a_1 \ldots a_n)$ is smaller than or equal to $up(a_1 \ldots a_m)$ for a prefix $a_1 \ldots a_m$ of $a_1 \ldots a_n$, then $up(a_1 \ldots a_n)$ is at least $up(a_1 \ldots a_m) + 1$.

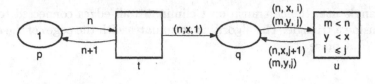

var n,m,x,y,i,j: nat

Figure 23.8. Longest upsequence

24   Exclusive Writing and Concurrent Reading

As a variant of the mutual exclusion algorithms of Sect. 13, here we consider
the case of conditional exclusive and concurrent access to a scarce resource,
e.g., a variable that may be updated only exclusively by one of its writer
processes, but be read concurrently by its reader processes.

24.1   An unfair solution

Figure 24.1 shows a first approach for an algorithm that organizes this version
of mutual exclusion: Any of a set W of initially *quiet* writer processes may
spontaneously get *pending* (quiescent action a), thus applying for a move
to *writing*. Likewise, each of a set R of initially *quiet* reader processes may
spontaneously get *waiting* (quiescent action d), thus applying for a move
to *reading*. There is a control token for each reader process which must be
available upon its move to *reading* (transition e). All such control tokens
must coincidently be available for a writer process to move from *pending* to

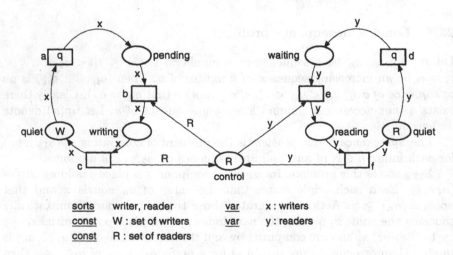

Figure 24.1. An unfair solution

writing (transition *b*). The control tokens in fact guarantee the required safety property: Whenever a writer process is writing, then no other writer process is writing and no reader process is reading.

However, *evolution*, as discussed in Sect. 13.1, is not guaranteed for writer processes, and further more cannot be achieved by the assumption of fairness.

24.2 A fair solution

Evolution has been achieved in $\Sigma_{24.2}$, with an additional synchronizing place, *key*, and the refinement of *pending* and of *waiting* into two consecutive places, respectively. *key* indicates that no writer process is at *pend2*, and *key* is a side condition for each reader process to move to *wait2* (with transition *f*). Three transitions, *b*, *f*, and *g* are assumed to be fair.

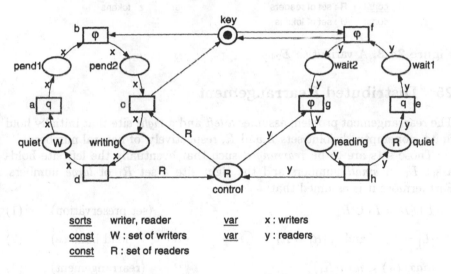

sorts	writer, reader	var	x : writers
const	W : set of writers	var	y : readers
const	R : set of readers		

Figure 24.2. Exclusive writing and concurrent reading

24.3 A variant of the solution

$\Sigma_{24.2}$ prevents competition among reader processes by a "private" control token for each of them. As a generalization we may assume a set U of control tokens, independent from the set R of reader processes (with $|U| < |R|$), such that each reader process must get hold of *any* such $u \in U$ in its step from *wait2* to *reading*. We furthermore may want to reduce the number of fair transitions. Figure 24.3 shows a solution with two fair transitions, *b* and *f*.

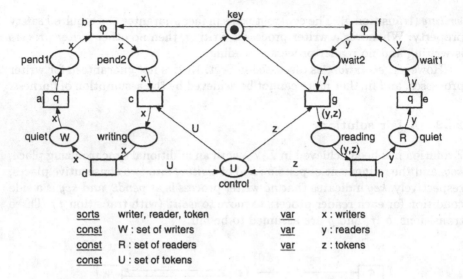

sorts	writer, reader, token	var	x : writers
const	W : set of writers	var	y : readers
const	R : set of readers	var	z : tokens
const	U : set of tokens		

Figure 24.3. A variant to $\Sigma_{24.2}$

25 Distributed Rearrangement

The rearrangement problem assumes a *left* and a *right* site that initially hold finite, nonempty, disjoint sets L and R, respectively, of natural numbers.

Those sets are to be *rearranged* such that eventually the left site holds a set L_1 of *small* numbers and the right site a set R_1 of *large* numbers. Furthermore it is assumed that:

$$L \cup R = L_1 \cup R_1 \qquad\qquad \text{(set preservation)} \qquad (1)$$

$$|L| = L_1 \quad \text{and} \quad |R| = R_1 \qquad\qquad \text{(load balance)} \qquad (2)$$

$$max(L_1) < min(R_1) \qquad\qquad \text{(rearrangement)} \qquad (3)$$

A distributed algorithm is to be constructed that does without additional storage for the two sites. Such an algorithm will be derived in the sequel, in a sequence of refinement steps.

25.1 First steps towards a solution

Figure 25.1 shows a first solution to this problem. This solution is not distributed, however, because occurrence of a requires data-dependent synchronization among the two sites; hence the algorithm is not really distributed. Data-dependent synchronization is avoided in the solution $\Sigma_{25.2}$, as *any* values stored at the two sites may engage in occurrences of a. But this algorithm is not guaranteed to terminate: If in a state s, the action a is enabled in mode m with $m(x) < m(y)$, the infinite sequence $s \xrightarrow{m} s \xrightarrow{m} \ldots$ is a feasonable interleaved run.

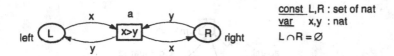

Figure 25.1. Non-distributed rearrangement

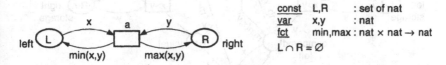

Figure 25.2. Non-terminating rearrangement

25.2   A handshake solution

The following step gives each site control over the next value to be offered for comparison: The *actual* places of $\Sigma_{25.3}$ always hold exactly one token, l and r, respectively, to be compared next or to be replaced by a "better" value. This algorithm still fails to terminate, but termination can be achieved if each comparison of values engages at least one "better" value. This is achieved in $\Sigma_{25.4}$: Each *compared* value is replaced by a better *actual* value from the respective storage. Comparison of values requires at least one newly chosen *actual* value. Hence the algorithm terminates in a state where no site has to offer a fresh actual value. $\Sigma_{25.4}$ is hence a perfect solution with handshake communication.

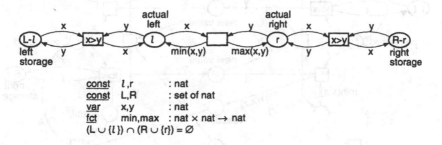

Figure 25.3. Rearrangement with distinguished candidates

25.3   A distributed solution

The handshake solution $\Sigma_{25.4}$ now serves as a basis for a distributed solution. To this end, each of the three communicating transitions is replaced by two message-passing transitions, as in $\Sigma_{25.5}$. This algorithm can apparently be

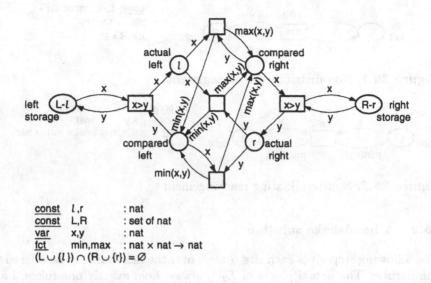

```
const    l ,r        : nat
const    L,R         : set of nat
var      x,y         : nat
fct      min,max     : nat × nat → nat
(L ∪ {l}) ∩ (R ∪ {r}) = ∅
```

Figure 25.4. Distributed handshake rearrangement

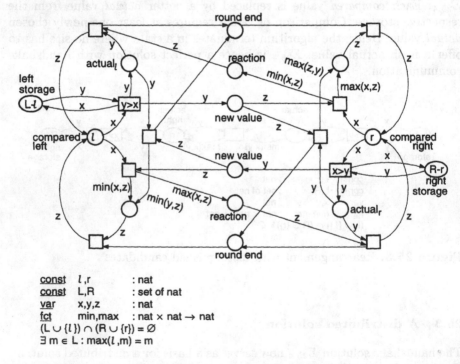

```
const    l ,r        : nat
const    L,R         : set of nat
var      x,y,z       : nat
fct      min,max     : nat × nat → nat
(L ∪ {l}) ∩ (R ∪ {r}) = ∅
∃ m ∈ L : max(l ,m) = m
```

Figure 25.5. Distributed message passing rearrangement

conceived as a variant of the crosstalk algorithm, $\Sigma_{12.5}$. The algorithm may work concurrently to the rise of the sets to be rearranged: There may be transitions that continuously drop new elements into the left and the right storage during the rearrangement operations. To start computation, assume at least one $m \in L$ with $\max(l, m) = m$. The symmetrical argument, at least one $m \in R$ with $\max(r, m) = r$, would suffice, too.

26   Self Stabilizing Mutual Exclusion

26.1   Self stabilization of mutual exclusion

A set of processes is assumed that include particular local states called *critical* states. A global state is said to guarantee *mutual exclusion* if at each of its reachable states, at most one process is *critical*. An algorithm is to be constructed which eventually leads to a state that guarantees mutual exclusion. As a particular difficulty, processes may occasionally execute *irregular* steps. Such a step may result in a state that does not guarantee mutual exclusion. The intended algorithm is supposed to be *self stabilizing* in this case, i.e., it should eventually lead to a state that again guarantees mutual exclusion.

In the sequel we solve this problem for sequences of tightly coupled, sequential processes.

26.2   Self stabilizing mutual exclusion for a sequence of four processes

A *stabilizing process* consists of four states, *critical*, *right*, *waiting*, and *left* that are visited in a circle, as in

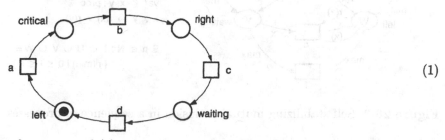

$$(1)$$

The four steps of (1) are called *regular* steps. Any other step between two different local steps is *irregular*; hence (1) exhibits eight irregular steps, not explicitly represented.

Now assume four stable processes, tightly coupled in a sequence as in Fig. 26.1. A stable process at *right* or at *left* is pending for a synchronized step with its right or left neighbor, respectively. A stable process is *waiting* until its right neighbor has reached its *left* state. A state $a \subseteq P_{\Sigma_{26.1}}$ is *feasible* if each stabilizing process contributes exactly one local state, i.e., a is formed

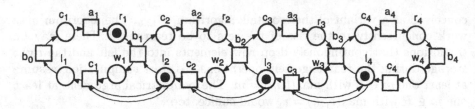

Figure 26.1. A sequence of four stable processes

$a = \{a_1, \ldots, a_4\}$ with $a_i \in \{c_i, r_i, w_i, l_i\}$ for $i = 1, \ldots, 4$. Each regular or irregular step, as defined above, retains feasibility of states.

Starting from *any* feasible state, $\Sigma_{26.1}$ eventually reaches the state $\{l_1, \ldots, l_4\}$. Mutual exclusion is guaranteed from then on, i.e., at most one process i is *critical* (i.e., at c_i) at each state that is reachable from $\{l_1, \ldots, l_4\}$.

Formal description and proof of those properties is postponed to Sect. 82.

26.3   Self stabilizing mutual exclusion for a sequence of processes

Figure 26.2 shows the self stabilizing mutex algorithm for any sequence of self stabilizing processes. Initially, some processes are *critical*, *waiting*, at *right*, or at *left*, respectively. Irregular steps are again not represented explicitly.

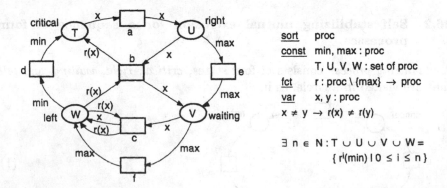

Figure 26.2. Self stabilizing mutual exclusion in a sequence of processes

V. Case Studies Continued:
Acknowledged Messages

In networks of communicating agents, the senders of messages frequently expect acknowledgments from their receivers: Transmission lines may be unreliable or the sender may prevent message overtaking or may wish to wait for further action until a set of messages has reached their respective destination.

We start with two *communication protocols*, i.e., distributed algorithms that detect and repair faulty transmission. Next we discuss algorithms that organize acknowledgments of messages to neighboring receivers in a network. Finally we consider the asymmetrical case of a master process that obtains acknowledgments or refusals from a set of slave processes.

27 The Alternating Bit Protocol

In a sequence of steps, a distributed algorithm will be derived that detects loss of messages and enforces transmission of copies of lost messages.

27.1 Unreliable transmission lines

A *communication protocol* establishes reliable message passing along unreliable transmission lines. There exist various forms of unreliability, including loss, change of order, or falsification of messages. This section will assume that messages may get lost, but are never falsified. Occasionally their order is assumed not to change. Only finitely many consecutive messages may get lost, however. Figure 27.1 outlines the assumptions described above.

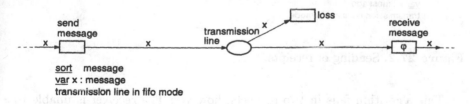

Figure 27.1. The transmission line

A *sender* and a *receiver* are assumed with actions *send messages* and *receive messages*, respectively. Fairness of *receive messages* excludes an infinite sequence of lost messages. Reliable message passing is guaranteed if an instance of each sent message will eventually reach its destination. The sender may repeat lost messages to this end, and the receiver may return receipts to the sender along another transmission line. Of course, this line may be unreliable, too.

In a sequence of steps, an algorithm will be derived that establishes reliable message passing along the unreliable transmission line of Fig. 27.1.

27.2 A first solution

For the sake of simplicity, in addition to the unreliable transmission line of Fig. 27.1, we temporarily assume a reliable transmission line from the receiver to the sender. As a very first idea, the receiver may acknowledge receipt of each message, as shown in Fig. 27.2. However loss of a message blocks the system. So, the sender may repeat a message, as in Fig. 27.3. In contrast to Sects. 16 and 19, in this section we allow *many identical* tokens at a place. In particular, the *transmission line* may hold several indistinguishable tokens.

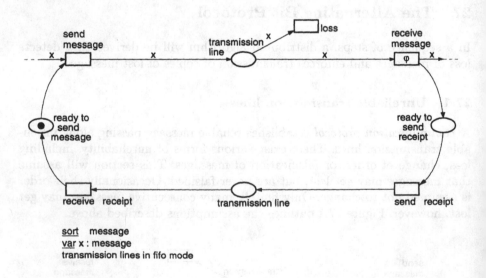

sort message
var x : message
transmission lines in fifo mode

Figure 27.2. Sending of receipts

This algorithm fails in two respects, however: The receiver is unable to distinguish a new, original message from copies of old messages, and the sender may entirely ignore the arrival of receipts, thus forever repeating copies of a message, instead of eventually receiving its receipt.

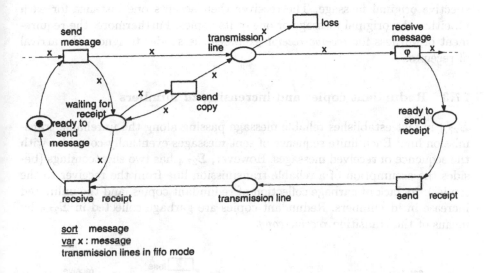

sort message
var x : message
transmission lines in fifo mode

Figure 27.3. Repetition of messages.

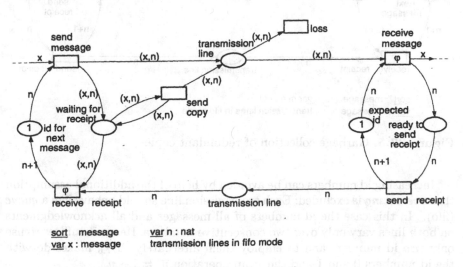

sort message   **var** n : nat
var x : message   transmission lines in fifo mode

Figure 27.4. Unique identification numbers (id)

Both problems have been overcome in $\Sigma_{27.4}$: Each message is given a unique identification number (id), with each copy assigned the id of the respective original message. The receiver then accepts one instance for each id, either the original message or one of its copies. Furthermore, the requirement of fairness for *receive receipt* excludes the sender to ignore the arrival of receipts.

27.3 Redundant copies and increasing *id* numbers

$\Sigma_{27.4}$ in fact establishes reliable message passing along the unreliable transmission line: Each finite sequence of sent messages eventually coincides with the sequence of received messages. However, $\Sigma_{27.4}$ has two shortcomings (besides the assumption of a reliable transmission line from the receiver to the sender): its lack of garbage collection of redundant copies, and the unlimited increase in id numbers. Redundant copies are garbage collected in $\Sigma_{27.5}$ by means of the transition *receive copy*.

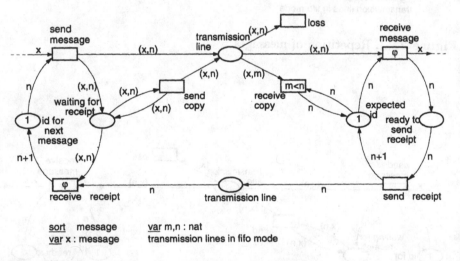

Figure 27.5. Garbage collection of redundant copies

Increasing id numbers can be avoided by help of the additional assumption that *overtaking* is excluded: Each transmission line should behave like a *queue* (fifo). In this case the id numbers of all messages and all acknowledgments on both lines vary only over two consecutive numbers. Hence it suffices to use only *two* id numbers and to employ them alternately, $\Sigma_{27.6}$ makes do with the id numbers 0 and 1 and the swap operation $\bar{n} := 1 - n$.

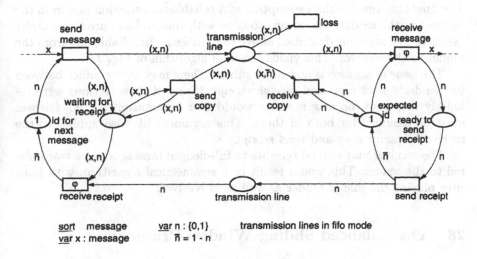

sort message   var n : {0,1}   transmission lines in fifo mode
var x : message   n̄ = 1 - n

Figure 27.6. Alternating identification numbers

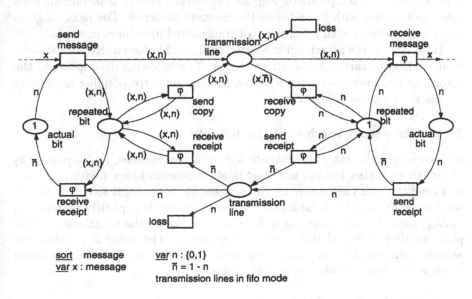

sort message   var n : {0,1}
var x : message   n̄ = 1 - n
transmission lines in fifo mode

Figure 27.7. The alternating bit protocol

27.4   The final solution

The final step revokes the assumption of a reliable transmission line from the receiver to the sender. The means to cope with this problem are structurally identical to the means described above to manage an unreliable line from the sender to the receiver. This yields the final algorithm of Fig. 27.7.

The issue of fairness is now more subtle: There may arise conflict between the sender's *send copy* and *receive receipt*. If one of those actions were infinitely neglected, no new message would ever be transmitted. So, fairness must be assumed for both of them. This argument likewise applies to the receiver's *receive copy* and *send receipt*.

The receiver may extend receipts to full-fledged messages to be transmitted to the sender. This would result in a symmetrical algorithm, with both sites playing the role of sender as well as of receiver.

28   The Balanced Sliding Window Protocol

The alternating bit protocol follows quite a strict policy: The sender receives a receipt for the i-th message before sending its $(i + 1)$st message. Here we consider a more liberal protocol that moves a "window" along the message sequence, consisting of two indices. Any data between both indices may be sent by the sender. The order of receipts likewise varies in a "window". As a first version of such a protocol, Fig. 28.1 shows the case of transmitting each message together with its index in the message sequence. The next diagram $\Sigma_{28.2}$ will make do with a finite set of transmitted identifiers instead.

Both sites of the algorithm are structurally and behaviorally almost identical, and both start in symmetrical states. The following description of the algorithm concentrates on the left site, leaving the corresponding arguments for the right site as an exercise.

28.1   The actual window of the left site

In each reachable state, the left site has its *actual window*, i.e., a pair (a, b) of indices such that the left site may freely choose an index i with $a < i \leq b$ and send the i-th message (i, x) (a *lr-message*) to the right site. The actual window (a, b) consists of the *lower window index* $a = k$, explicitly represented in $\Sigma_{28.1}$, and the *upper window index* $b = j + w$, with j the actual value of the place *smallest index of still expected rl-messages*. The value w is called the *window constant*, i.e., an integer constant of the system. Transition a hence sends *lr-messages* within the actual window.

28.2   Window bounds of the left site

With a message (i, x) received by the left site, the right site acknowledges receipt of all messages with indices from 1 up to $i - w$. Hence the left site

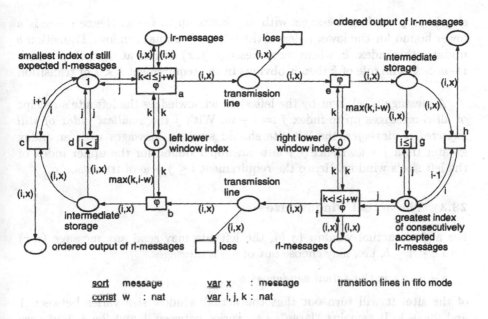

Figure 28.1. Balanced sliding window protocol with unbounded indices

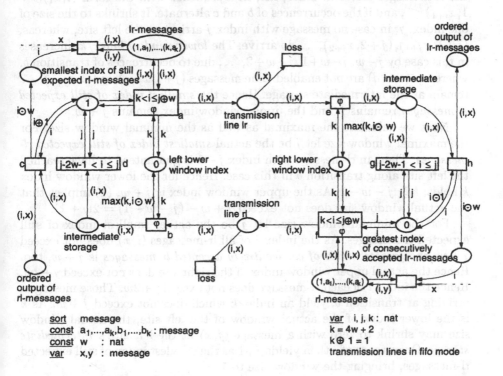

Figure 28.2. Balanced sliding window protocol with bounded indices

should stop sending messages with any index up to $i - w$. Hence $i - w$ is a lower bound for the lower index of the left site's actual window. Transition b updates this index, k, whenever a message (i, x) arrives at the left site with $i - w > k$. The role of k then is obvious in the requirement $k < i$ of transition a.

A message (i, x), sent by the left site, acknowledges the left site's receipt of all rl-messages up to index $j := i - w$. With j the *smallest index of still expected rl-messages*, the left site should send no messages with an index greater than $j + w$. Hence $j + w$ is an upper bound for the upper index of the left site's window. Hence the requirement $i \leq j + w$ of transition a.

28.3   The actual window size

For a given actual window (a, b), the left site may send any message (i, x) with $a < i \leq b$, i.e., may choose out of $b - a$ messages.

$$b - a \qquad \text{is the } actual\ window\ size \tag{1}$$

of the site. It will turn out that the actual window size varies between 1 and $2w + 1$. It remains "large", i.e., varies between 1 and $2w + 1$ in case the rl-messages arrive in order, i.e., in a sequence with the form $(i, x_i)(i + 1, x_{i+1}) \ldots$, and if the occurrences of b and c alternate. It shrinks to the size of one index, j, in case no message with index j arrives at the left site, whereas, $(j + 1, x_{j+1}), (j + 2, x_{j+2}), \ldots$ do arrive: The *lower window index*, k, increases in this case by $j - w, j - w + 1, j - w + 2, \ldots$, due to occurrences of transition b, whereas c (and d) are not enabled. The messages $(j+1, x_{j+1}), (j+2, x_{j+2}), \ldots$ remain at the intermediate storage. Hence the *smallest index of still expected rl-messages* remains j, and the upper window index remains $j + w$.

Next we calculate the maximal as well as the minimal window size: For the maximal window size let j be the actual *smallest index of still expected rl-messages*. Then an rl-message with index $j - 1$ is guaranteed to have reached the left site along transition b in this case. Hence for the lower window index k holds: $k \geq j - w - 1$. As the upper window index is $j + w$, (1) implies that the actual window size does not exceed $j + w - (j - w - 1) = 2w + 1$.

For the minimal window size let j be the actual smallest index of still expected rl-messages. As the index i of all lr-messages (i, x) does not exceed $j + w$, the *greatest index of consecutively accepted lr-messages* is $j + w$, too. Hence the actual upper window index of the right site does not exceed $j + 2w$. Hence the index i of each rl-message does not exceed $j + 2w$. Those messages, arriving at transition b, yield an index k which does not exceed $j + w$. As k is the lower index of the actual window of the left site, the actual window size may shrink to zero with a message (j, x) at the left site's *intermediate storage*. Occurrence of c then yields $j + 1$ as the smallest index of still expected rl-messages, bringing the window size to 1.

28.4   The right site's window

The two sites essentially differ only in one aspect: The action c increases the *smallest index of still expected rl-messages*, whereas the action h does not increase the *greatest index of consecutively accepted lr-messages*. However, this difference does not really affect the site's behavior, given a slight adjustment at the guards of transitions f and g.

28.5   Bounded indices

The above version $\Sigma_{28.1}$ of the sliding window works with a strictly increasing sequence of indices. A finite set of indices, applied in cyclic order, is sufficient, however.

To this end we assume – by analogy to the alternating bit protocol – that overtaking is excluded: Each transmission line should behave as a queue. Secondly we have to estimate the number of different messages that coincidently may exist in the system. This number essentially depends on the window constant, w. With j the *smallest index of still expected rl-messages*, the messages in the lr transmission line may be indexed from $j - w - 1$ to $j + w$, according to the considerations of Sect. 28.3. Messages at the right site's *intermediate storage* and at the *greatest index of consecutively accepted messages* vary in the same range. Due to the guards of transition f, messages in the rl transmission line are indexed between $j - 2w - 1$ and $j + 2w$. A message (i, x) in the left site's intermediate store is a copy of a previously received message if i varies between $j - 2w - 1$ and j. Hence the guard of transition d. The guard of transition g follows the same line of arguments. Altogether, including both limits, there may be messages around with up to $4w + 2$ different indices. Hence it suffices to employ $4w + 2$ identification numbers and to employ them in cyclic order. $\Sigma_{28.2}$ uses the id numbers $1, \ldots, 4w + 2$.

28.6   Specializations and generalizations

The alternating bit protocol is essentially a special case of the sliding window protocol, with window constant $w = 0$. *Receive message* and *receive copy* of $\Sigma_{27.7}$ then correspond to the transitions h and g of $\Sigma_{28.2}$. Transitions e and f of $\Sigma_{28.2}$ just organize the window's slide.

As a generalization, each site may employ its own window constant. The actual window size may shrink to zero or to some number $k > 1$ in this case. This requires more subtle fairness assumptions.

29   Acknowledged Messages to Neighbors in Networks

A *(distributed) network* includes a finite set of *sites*. Two sites may be *neighbors*, i.e., be linked by a *transmission line* for messages to be transmitted in either direction. A site may send messages to its neighbors and expect receipts that acknowledge the messages. This section presents some aspects of algorithms that organize acknowledged messages to neighbors in networks.

29.1   One sender

Let i be a site in a network (the *initiator*) and let U be the set of its neighbors. Figure 29.1 shows the basics of an acknowledged message sent by i to all its neighbors. After sending the message to each neighbor (action a), the initiator remains waiting until receipts have arrived from all neighbors. Then the initiator terminates (action b). Each single uninformed neighbor $x \in U$ receives the message (action c) and returns a receipt (action d). The algorithm likewise works in a round-based version, as in Fig. 29.2.

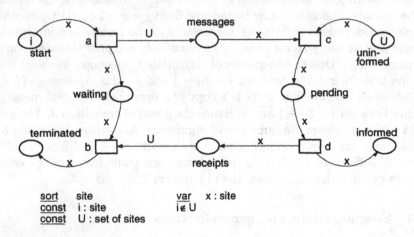

<pre>
sort   site   var   x : site
const   i : site   i∉ U
const   U : set of sites
</pre>

Figure 29.1. Basics of acknowledged messages to neighbors

29.2   Many senders

Matters are more involved in the case of more than one initiator: Each message and each receipt must include its target as well as its source. In Fig. 29.3, messages and receipts are represented as pairs (target, source). For each site $u \in U$, $pr_1(N(u))(= pr_2(\overline{N}(u)))$ is the set of neighbors of u. Furthermore, $N(u)$ and $\overline{N}(u)$ are the sets of messages sent by u and receipts received by u, respectively.

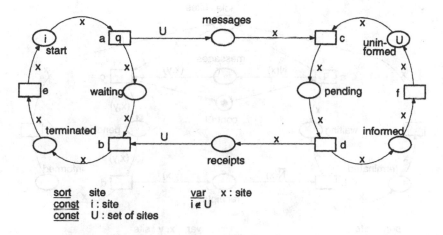

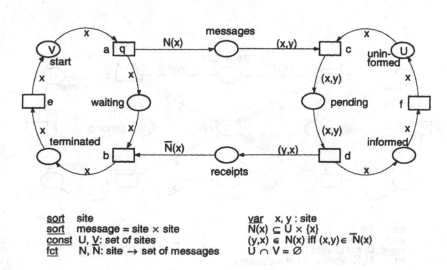

Figure 29.2. Round-based message passing

Figure 29.3. Message passing by many initiators

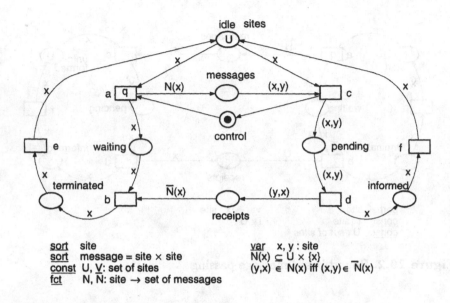

Figure 29.4. Sites acting as sender or as receiver

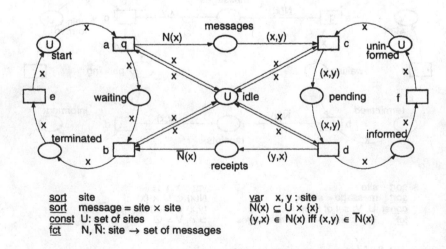

Figure 29.5. Sites acting as sender and as receiver

29.3   Variants

As a variant, a site may decide to act either as a sender or as a receiver of messages, as in Fig. 29.4. This algorithm would deadlock if more than one site could act as a sender at the same time.

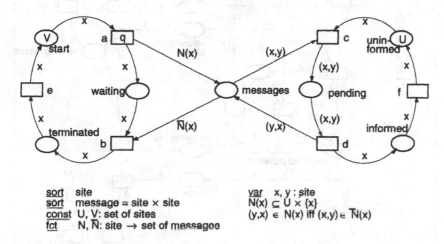

sort   site
sort   message = site × site
const U, V: set of sites
fct   N, N̄: site → set of messages

var   x, y : site
N(x) ⊆ U × {x}
(y,x) ∈ N(x) iff (x,y) ∈ N̄(x)

Figure 29.6. Joint messages and receipts

As a further variant, each site may act as a sender and as a receiver. This can easily be achieved: In Fig. 29.3 replace the requirement $U \cap V = \emptyset$ by $U = V$. Each site then consists of two independent, concurrently acting components: a sender and a receiver. One may replace them by one sequential, nondeterministic component, as in Fig. 29.5.

Finally we observe that the distinction of messages and receipts is superficial, as they are clearly identified by their respective source and target. Both will be called *messages* in the sequel. In Fig. 29.6, the place *messages* includes all messages that have been sent but not yet received.

30   Distributed Master/Slave Agreement

A particular form of message acknowledgment occurs in the following protocol: Assume a "master" process and a set U of "slave" processes. Update orders launched by the master are to be executed by the slaves, provided no slave refuses. In order to achieve this behavior, the master first sends an inquiry to each slave. Each slave checks the inquiry and reports acceptance or refusal to the master. In case all slaves accept, the master sends an update order to each slave. In case one slave refuses, the master sends a cancellation to each slave.

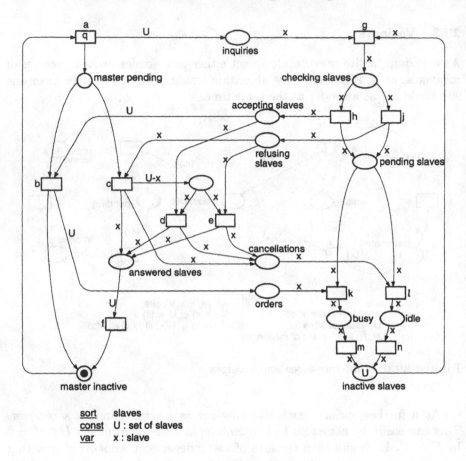

Figure 30.1. Distributed master/slave agreement

Figure 30.1 shows an algorithm that organizes this behavior. Initially, the master and all slaves are *inactive*, and the only activated transition is the quiescent transition a. Its occurrence starts the agreement procedure by the master's sending of *inquiries* to all slaves. Then each slave x on its own turns *checking* (action $g(x)$) and nondeterministically chooses to *accept* or to *refuse* the master's inquiry (action $h(x)$ or $j(x)$). In case all slaves $x \in U$ are *accepting*, the master sends *orders* to all slaves (action b). Otherwise at least one slave x is *refusing* and thus enables $c(x)$. Then each other slave y is notified to *cancel*: either by $d(y)$ (in case y had *accepted* the master's offer) or by $e(y)$ (in case y had *refused*). Eventually, all slaves $x \in U$ have *answered* and are sent *cancellations*. Altogether, all slaves x are forwarded either *orders*, or *cancellations* and so they all turn either *busy* or *idle* (action $k(x)$ or $l(x)$, respectively).

Hence the algorithm guarantees that *master pending* is eventually followed by *master inactive* together with either all slaves *busy* or all slaves *idle*. This property will formally be shown in Part D.

VI. Case Studies Continued: Network Algorithms

A distributed algorithm is said to be a *network algorithm* if it is not intended to run on just one fixed network. Rather, a network algorithm is a *schema* of algorithms, to run on any in a whole class of networks, such as the connected networks, the ring- or tree-shaped networks, etc.

Network algorithms have many features in common, and it is quite convenient to represent equal features always alike. Some intuition-guided conventions and principles for the representation of network algorithms will be presented in this chapter. They have already been employed in the above algorithms, and will likewise be used in all algorithms of this chapter, including algorithms for mutual exclusion, consensus, and self-stabilization in networks.

31 Principles of Network Algorithms

The fundamental idea of the representation of network algorithms is the *generic representation* of local algorithms, and the explicit representation of messages under way. This implies a canonical representation of network algorithms, according to the *locality principle* and the *message principle*.

31.1 Generic local algorithms

Many sites of a network usually run the same local algorithm. A network algorithm usually consists of a few, up to about three, different local algorithms. In a system net representation, each local algorithm is represented *generically*, with a variable denoting the network sites. All local algorithms are connected to a place, usually called *messages*, that includes all messages already sent by their source site and not yet received by their target site.

In technical terms, a network algorithm is represented as a net schema. Each local algorithm employs a variable (usually x) for the active site. Each action is supposed to be executed by the site x. The following *locality principle* guarantees that each action employs only data that are local to x:

31.2 The locality principle

For each transition t, each ingoing arc of t is inscribed by a set of n-tuples of variables (mostly just one n-tuple, often a pair or even just a single variable). The first variable of all n-tuples of all ingoing arcs of t are identical (usually x).

In fact, all nets in Sects. 29 and 33 follow the locality principle, with the exception of $\Sigma_{29.4}$: Transition a has a dot inscribed ingoing arc. In fact, this algorithm is not a network algorithm due to the place $control$: All sites may compete for its token. Hence, $control$ does not establish communication between just two sites.

The following principle of message representation is an offspring of the above locality principle.

31.3 The message principle

Each message is represented as a n-tuple $(x_1, \ldots, x_n)$ with x_1 the receiver and x_2 the sender of the message. $x_3, \ldots, x_n$ may contain any kind of information. (The case of $n = 2$ is quite frequent).

In fact this principle has been applied throughout Sects. 27 and 30, and will likewise be followed in forthcoming sections.

Summing up, the above representation rules provide a syntactical criterion for the distributedness of an algorithm. It supports clarity and readability of network algorithms, including a standard representation of messages.

31.4 Some notions, notations, and conventions

As usual, for a set U and a relation $W \subseteq U \times U$, let uWv iff $(u, v) \in W$. Furthermore,

  i. $W_1 := \{u \in U \mid$ ex. $v \in U$ with $uWv\}$,
     $W_2 = \{v \in U \mid$ ex. $u \in U$ with $uWv\}$
 ii. $W(u) := \{v \in U \mid uWv\}$
iii. $W^{-1} := \{(v, u) \mid uWv\}$ (frequently written $\overline{W}$)
 iv. uW^+v iff for some $n \geq 1$ and some $u_0, \ldots, u_n \in U$, $u_0 = u$, $u_n = v$, and
     $u_0 W u_1 \ldots u_{n-1} W u_n$
  v. uW^*v iff uW^+v or $u = v$.

The forthcoming system schemata all assume any underlying $network$. In an abstract, technical setting, a network is a $graph$; it will usually be described by its sets U of $nodes$ and W of $arcs$. Each arc is a pair of nodes. $W(x)$ denotes the set of $neighbors$ of a node x. The network is frequently $symmetrical$ $(W = W^{-1})$ and $connected$ $(xW^*y$ for all $x, y \in U)$. W usually covers exactly the nodes of U $(W_1 \cup W_2 = U)$.

W is a *tree with root* u iff each node is reachable from u ($\forall x \in U : uW^*y$), W is cycle free ($xW^+y \to x \neq y$), and each node has at most one predecessor ($yWx \wedge zWx \to y = z$).

W is an *undirected tree* iff W is symmetrical, connected, and no undirected sequence of arcs forms a cycle ($x_0Wx_1 \ldots x_nWx_{n+1} \wedge x_{i-1} \neq x_{i+1}$ ($i = 1, \ldots, n$) $\to x_0 \neq x_n$).

32 Leader Election and Spanning Trees

32.1 The basic leader election algorithm

The sites of a network are frequently supposed to elect one site as their *leader*. In case the leader site crashes, a new leader must be elected. The sites are given unique names to this end (e.g., integer numbers) and a total order is assumed on those names.

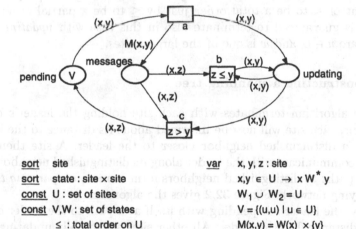

<u>sort</u> site	<u>var</u> x, y, z : site
<u>sort</u> state : site × site	x,y $\in$ U $\to$ x W* y
<u>const</u> U : set of sites	W$_1$ $\cup$ W$_2$ = U
<u>const</u> V,W : set of states	V = {(u,u)	u $\in$ U}
$\leq$: total order on U	M(x,y) = W(x) × {y}
<u>fct</u> M : state $\to$ set of states	

Figure 32.1. Basic leader election

Figure 32.1 gives a distributed algorithm for the election of a leader in any connected network. Initially, each site is *pending* and assumes its own name as a candidate for the leader. In later states, a *pending* site holds a better candidate, i.e., one with a larger name. Generally, a *pending* site u together with its actual candidate v is represented as a state (u,v). Upon *pending* with v, u informs each neighbor in $W(u)$ about v by action $a(u,v)$ and then becomes *updating*. An *updating* site u with its actual leader candidate v may receive a *message* (u,w). In case the newly suggested candidate, w, does not exceed v, the site u remains *updating* with v (action $b(u,v,w)$). Otherwise u

goes *pending* with the new candidate w (action $c(u, v, w)$) and continues as described above.

A message $(w, v) \in M(u, v)$ takes the form of a state, with u informing the site w about v as a candidate for the leader. There may occur multiple copies of identical messages (as in case of communication protocols). This can easily be fixed, by extending each message with its sender.

$\Sigma_{32.1}$ does not perfectly meet the message principle Sect. 31: A message (u, v) in $\Sigma_{32.1}$ consists of its receiver u and a further piece of information, v. The sender is not mentioned explicitly (though it was easy to do so).

Given a connected network with a finite set U of sites and a total order $\leq$ on U, the algorithm terminates with *updating* all pairs (u, w), where $u \in U$ and w is the maximal element of U.

32.2   A variant of the basic algorithm

In the more general case of a partial order each site may select one of the largest sites as its leader. This is easily achieved: In Fig. 32.1, replace the requirement of $\leq$ to be a total order just by $\leq$ to be a partial order. The algorithm is guaranteed to terminate also in this case with *updating* pairs (u, w), where $u \in U$ and w is one of the largest sites.

32.3   Constructing a spanning tree

The above algorithm terminates with each site holding the leader's name. As a variant, each site will now be informed about its distance to the leader and about a distinguished neighbor closer to the leader. A site then may effectively communicate with the leader along its distinguished neighbor. The respective paths to distinguished neighbors form a *minimal spanning tree* in the underlying network. Figure 32.2 gives the algorithm.

Initially, the leader r is pending with itself as a path to the leader candidate, and distance 0 to the leader. All other sites are initially updating with the unspecified leader candidate $\perp$ and infinite distance. In later phases, a *pending* token (u, v, n) indicates that there is a path of length n from u along v to the leader. A *pending* site u forwards its actual distance n to all its neighbors (by action $a(u, v, n)$) and then turns *updating*. An *updating* token (u, v, n) may receive a message (u, w, m). In case the reported distance m of w to the leader would not improve the actual distance n, the site u remains with distance n along neighbor v (action $b(u, v, w, n, m)$, with ordered set (x, y, z, i, j) of variables). Otherwise u goes pending with distance $m + 1$ along neighbor w (action $c(u, v, w, n, m)$, with ordered set (x, y, z, i, j) of variables).

This algorithm can be generalized to a set $R \subseteq U$ of leaders in the obvious way: Initially, *pending* carries $\{(r, r, 0) \mid r \in R\}$ and *updating* $\{(u, \perp, \omega) \mid u \in U \setminus R\}$. The algorithm then terminates with *updating* triples (u, v, n), where

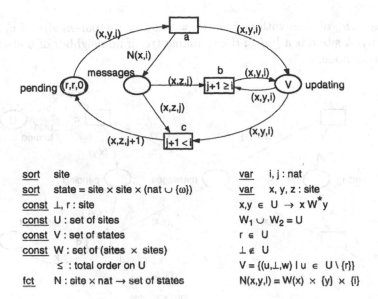

<u>sort</u> site	<u>var</u> i, j : nat
<u>sort</u> state = site × site × (nat ∪ {ω})	<u>var</u> x, y, z : site
<u>const</u> ⊥, r : site	x,y ∈ U → x W* y
<u>const</u> U : set of sites	W₁ ∪ W₂ = U
<u>const</u> V : set of states	r ∈ U
<u>const</u> W : set of (sites × sites)	⊥ ∉ U
≤ : total order on U	V = {(u,⊥,w)	u ∈ U \ {r}}
<u>fct</u> N : site × nat → set of states	N(x,y,l) = W(x) × {y} × {i}

Figure 32.2. Shortest distance to a root

n is the minimal distance to a leader and v the name of a neighbor closer to
a leader.

33   The Echo Algorithm

Given a finite, connected network with a particular initiator site, the echo algorithm organizes acknowledged broadcast of the initiator's message throughout the entire network to all sites: The initiator will terminate only after all
other sites are informed.

33.1   One initiator in one round

Figure 33.1 shows one round of messages, sent by the initiator i to all its
neighbors, just as in Fig. 29.1. Furthermore, messages and receipts are jointly
represented in one place, in accordance with Fig. 29.6. The central idea of the
echo algorithm is now covered in the step from $\Sigma_{33.1}$ to $\Sigma_{33.2}$: Upon receiving
the initiator's message, a neighbor of the initiator forwards the message to all
its neighbors but the initiator, and remains pending until receiving messages
from all those neighbors. Each site is eventually addressed in this schema.
Each uninformed site $u \in U$ receives in general more than one message, hence
u selects one occurrence mode (u, v) of action c. In this case, v is called the
parent site of u. The pairs (u, v) with v the parent site of u, form a *spanning
tree* in the underlying network: For each site $u \in U$ there exists a unique

sequence $u_0 \ldots u_n$ of sites with $u_0 = u$, $u_n = i$ and u_i the parent site of u_{i-1} $(i = 1, \ldots, n)$. A site u is a *leaf* of the spanning tree if no neighbor of u elects u as its parent node.

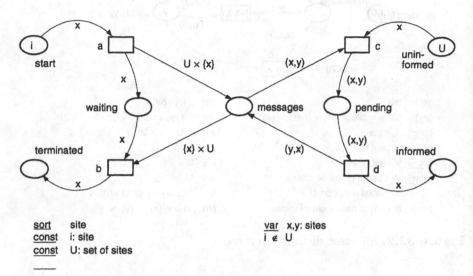

sort   site
const i: site
const U: set of sites

var x,y: sites
i $\notin$ U

Figure 33.1. The initiator informs its neighbors

For each pending leaf (u, v), the place *messages* eventually holds all messages $\overline{M}(u) - (u, v)$, hence the leaf becomes *informed* by occurrence of d in mode (u, v). The leaves are the first to become (concurrently) *informed*. Then all sites are consecutively *informed*, causally ordered along the spanning tree. Finally, the initiator's transition b is enabled, and the *waiting* initiator turns *terminated*.

33.2   One initiator in many rounds

The above one round echo algorithm likewise works also in a cyclic environment, as in Fig. 33.3.

33.3   Many initiators

Matters are more involved in the case of more than one initiator: The initiator's identity must be forwarded together with each message. Hence in $\Sigma_{33.4}$, each message is a triple (x, y, z) with receiver x, sender y and initiator z. A message (x, y, z) is sent by an initiator z if $y = z$ and is received by an initiator z if $x = z$. All non-initiators just forward the third component of messages.

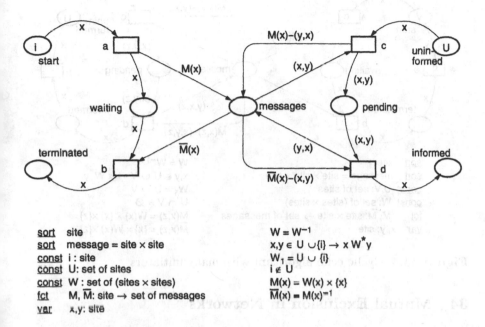

Figure 33.2. One-round echo algorithm

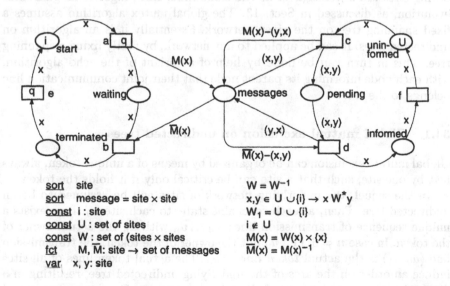

Figure 33.3. Cyclic echo algorithm

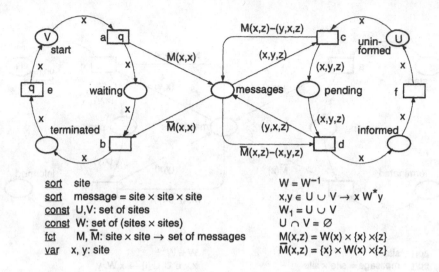

<u>sort</u> site	$W = W^{-1}$
<u>sort</u> message = site × site × site	$x,y \in U \cup V \to x\,W^*y$
<u>const</u> U,V: set of sites	$W_1 = U \cup V$
<u>const</u> W: set of (sites × sites)	$U \cap V = \varnothing$
<u>fct</u> M, $\overline{M}$: site × site → set of messages	$M(x,z) = W(x) \times \{x\} \times\{z\}$
<u>var</u> x, y: site	$\overline{M}(x,z) = \{x\} \times W(x) \times\{z\}$

Figure 33.4. Cyclic echo algorithm with many initiators

34   Mutual Exclusion in Networks

Two algorithms will be discussed in this section. The first algorithm guarantees *global* mutual exclusion: In the entire network, at most one site is critical at each time. The second algorithm guarantees *local* mutual exclusion: Neighboring sites are never critical at the same time. Both algorithms guarantee evolution, as discussed in Sect. 13. The global mutex algorithm assumes a fixed spanning tree on the given network. Essentially it is an algorithm on undirected trees. It can be applied to any network, by firstly fixing a spanning tree. This in turn can be done by help of a variant of the echo algorithm, with each node informing its parent node that their joint communication line belongs to the tree.

34.1   Global mutual exclusion on undirected trees

Global mutual exclusion can be organized by means of a unique token, always helt by one site, such that a site can be critical only if it holds the token.

In the sequel, the underlying network of sites will be assumed to be an undirected tree. Then, at each reachable state, to each site u_0 there exists a unique sequence of transmission lines, $u_0 \ldots u_n$ with u_n the actual owner of the token. In case $n \neq 0$, i.e., u_0 not the owner of the token, the transmission line (u_0, u_1) is the actual *token line* of u_0. The actual token lines of all sites induce an order on the arcs of the underlying undirected tree, resulting in a directed tree with the owner of the token as its root. $\Sigma_{34.1}$ organizes global mutual exclusion on trees: The place *token* holds the actual owner of the token; N is a directed tree on the sites of the network, such that the actual

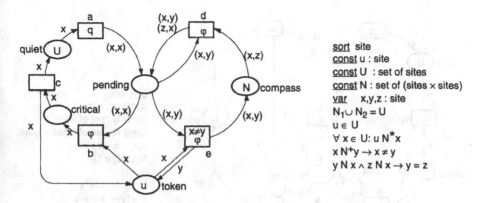

Figure 34.1. Global mutex on undirected trees

owner of the token is the root of the tree. A *quiet* site u may strive for the token by occurrence of $a(u)$ and then go *critical* by occurrence of $b(u)$, provided u presently owns the token.

If u does not own the token, either *compass* holds the token line (u,v) of u, or the reverse (v,u) of the token line is *pending*. Intuitively formulated, (u,v) at *compass* states that in order to obtain the token, u must send a corresponding request to v, by occurrence of $d(u,v)$. A *pending* token (u,v) states that u has the duty to get hold of the token and to hand it over to v. If u holds the token already, u hands it over to v by $e(u,v)$. Otherwise u has a token line, (u,w), at *compass*, and u sends a request for the token to w, by $d(u,v)$.

Three competing transitions, b, d, and e, are assumed to be fair in $\Sigma_{34.1}$.

34.2   A version with a simple fairness requirement

Figure 34.2 shows a variant of $\Sigma_{34.1}$ that requires only two transitions to be treated fairly. To this end, the place *pending* of $\Sigma_{34.1}$ has been refined into the sequence of place *job*, action f, and place *serving* in $\Sigma_{34.2}$. Each site u is concurrently *serving* at most one neighbor site, due to the place *idle*.

The essential difference between $\Sigma_{34.1}$ and $\Sigma_{34.2}$ is obvious whenever several sites, $v_1, \ldots, v_n$, say, are requesting the token from the same site, u. In $\Sigma_{34.1}$ this is represented by n tokens $(u,v_1), \ldots, (u,v_n)$ at *pending*. With (u,w) at *compass*, some v_j causes u to demand the token from w, by $d(v_j,u,w)$. After eventually having obtained the token, u selects a site v out of $v_1, \ldots, v_n$ and hands the token over to v (by $e(u,v)$), in case $v \neq u$.

In $\Sigma_{34.2}$ only one request, (u,v), is *serving* whereas all other pending requests, (u,v_i) with $v_i \neq v$, are at *job*. With (u,v) at *compass*, v will demand the token from w, by $d(v,u,w)$. After eventually having obtained the token, u hands it over to v, by $e(u,v)$, or goes critical by $b(u)$ in case $v = u$.

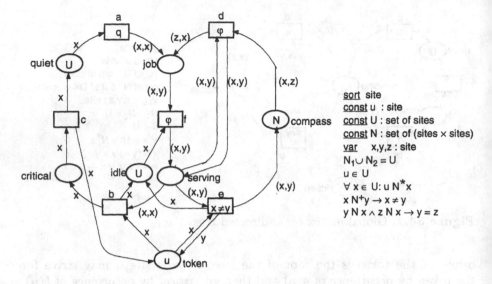

Figure 34.2. A version with a simple fairness requirement

34.3 Local mutual exclusion on networks

Here we consider networks without any restriction on their topology. By analogy to $\Sigma_{13.1}$, each site is assumed to be bound to the cyclic visit of essentially three local states, called *quiet*, *pending*, and *critical*, with a quiescent step from *quiet* to *pending*. In $\Sigma_{34.3}$, *pending* is refined to two states, *pend1* and *pend2*. Two sites are neighboring in the network if and only if they share a scarce resource. Each resource is shared by two sites.

An algorithm on a network with this kind of site guarantees *local mutual exclusion* iff neighboring sites are never both critical at the same time. It guarantees *evolution* iff each pending site will eventually be critical.

As a special case, the system $\Sigma_{10.1}$ of thinking and eating philosophers guarantees local mutual exclusion (with *eating* the respective critical state). However, this algorithm neither guarantees evolution, nor is it distributed.

Figure 34.3 shows a distributed algorithm that guarantees local mutual exclusion and evolution on networks. A resource shared by two sites u and v is represented by (u, v) or (v, u). Each resource at any time is on hand (though not necessarily in use) of one of its users. According to the locality principle (Sect. 31.2), (u, v) indicates that the resource shared between u and v is presently on hand at u. Occurrence of the quiescent action $a(u)$ indicates that the site u is about to get *critical*, in analogy to the actions a_l and a_r of Fig. 13.1. The step from *pend1* to *pend2* (action $b(u)$) demands $r(u)$ at *ready*, i.e., that u has re-organized all its resources after its last visit of *critical*. Details on this issue follow later. The crucial step of a site u, from *pend2* to *critical* (action $c(u)$) requires the set $r(u)$ of all resources of u to be

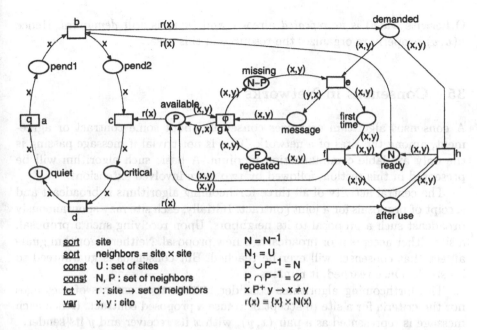

Figure 34.3. Local mutual exclusion on networks

available to u. Upon returning back to *quiet* by $d(u)$, each resource (u,v) is to be re-organized along *after use*.

Each resource that u shares with some neighbor v is in one of three states:

i. (u,v) is in *repeated*. In this case, the resource is on hand at u, and u has been its last user. Upon request of v by *message* (u,v), the site u may hand the resource over to v, by action $g(u,v)$.

ii. (u,v) is *missing*. In this case, the resource is on hand at v. In case u demands the resource (u,v), u sends a *message* (v,u) to v by action $e(u,v)$ and will eventually obtain the resource, by action $g(v,u)$.

iii. (u,v) is in *first time*. In this case, u will eventually obtain the resource from v and will not hand it over to v again before having been *critical* at least once.

A site u goes *critical* by occurrence of $c(u)$. This requires $r(u)$ be available to u. A resource (u,v) in *repeated* may be available to u, but u may decide to hand it over to v, by $g(u,v)$. For each resource (u,v) not available, u has previously sent a message (v,u) to v, by $b(u)$ and $e(u,v)$, and v will eventually hand over (u,v) to u, by $g(v,u)$. The resource (u,v) is at *first time* in this case. The site u retains all forks at *first time* after having been critical.

Each resource that a *critical* site u shares with a neighbor v is either freshly handed over to u, i.e., (u,v) is at *first time* or u has used it before already, i.e., (u,v) is at *repeated*. This implies two different actions for (u,v) at *after use*: In case of *first time*, $f(u,v)$ will occur and bring (u,v) to *repeated*.

Otherwise, (u, v) is at *repeated* already and (u, v) is still *demanded*. Hence $h(u, v)$ properly re-organizes the resource in this case.

35   Consensus in Networks

A consensus algorithm organizes consensus about some contract or agreement, among the sites of a network. This is not trivial if message passing is the only available communication medium. A basic such algorithm will be presented in this section, followed by two more involved extensions.

The central activity of all three forthcoming algorithms is broadcast and receipt of proposals for a joint contract. Initially, each site may spontaneously broadcast such a proposal to its neighbors. Upon receiving such a proposal, a site either accepts it or broadcasts a new proposal. Neither algorithm guarantees that consensus will ever be reached. But consensus is guaranteed to be *stable*: Once reached, it remains.

The forthcoming algorithms consider neither the contents of messages nor the criteria for a site to accept or refuse a proposed contract. Hence each message is represented as a pair (x, y), with x its receiver and y its sender.

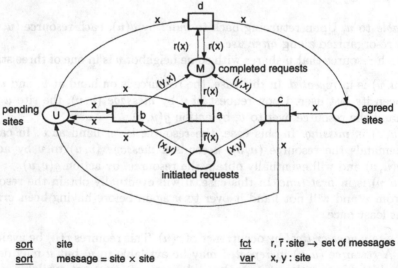

<u>sort</u>	site	<u>fct</u>	r, r̄ :site → set of messages
<u>sort</u>	message = site × site	<u>var</u>	x, y : site
<u>const</u>	U : set of sites		r(x) = {x} × M(x)
<u>const</u>	M : set of messages		r̄(x) = M(x) ×{x}

Figure 35.1. Basic algorithm for distributed consensus

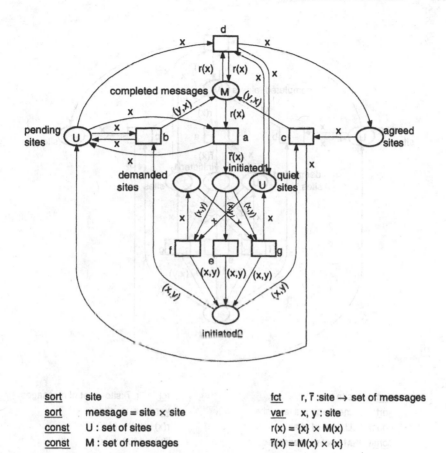

Figure 35.2. Distributed consensus with demanded negotiators

<u>sort</u>	site		<u>fct</u>	r, $\bar{r}$:site $\to$ set of messages
<u>sort</u>	message = site $\times$ site		<u>var</u>	x, y : site
<u>const</u>	U : set of sites			$r(x) = \{x\} \times M(x)$
<u>const</u>	M : set of messages			$\bar{r}(x) = M(x) \times \{x\}$

35.1 A basic consensus algorithm

Figure 35.1 shows an algorithm that organizes consensus. Initially, each site is *pending* and each message is *completed* (i.e., in the hands of its sender).

In this situation, any site x may send each neighbor y a message (y, x) (action $a(x)$). Upon receiving a message, a site x reads its contents and returns it to its sender y, by action $b(x, y)$ or action $c(x, y)$. Both actions $b(x, y)$ and $c(x, y)$ furthermore make the receiver x *pending*. Finally, each *pending* site x may turn *agreed*, provided all its messages $r(x)$ are *completed* (action $d(x)$).

Obviously, at any time, a site is either *pending* or *agreed*, and a message is either *completed* or *initiated*. The algorithm does not guarantee that the sites eventually are all agreed. However, the algorithm guarantees *stability*: It is terminated if and only if all sites are agreed.

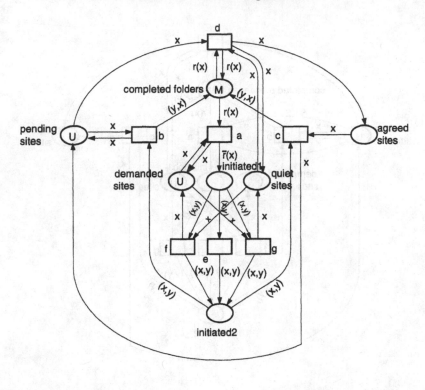

<u>sort</u> site	<u>fct</u> r, r̄ :site→set of messages
<u>sort</u> message = site × site	<u>var</u> x, y : site
<u>const</u> U : set of sites	r(x) = {x} × M(x)
<u>const</u> M : set of messages	r̄(x) = M(x) × {x}

Figure 35.3. A variant of $\Sigma_{35.2}$

35.2 An advanced consensus algorithm

The system $\Sigma_{35.2}$ extends $\Sigma_{35.1}$ by two further states, *demanded sites* and *quiet sites*. All sites are initially *quiet*. Each newly sent message (x, y) at the place *initiated*₁ may cause its receiver x to swap from *demanded* to *quiet* or vice versa. This is technically implemented in $\Sigma_{35.2}$ as a nondeterministic choice between actions $e(x, y)$ and $g(x, y)$, or $e(x, y)$ and $f(x, y)$, respectively.

A demanded site u is not *quiet*. If *demanded* and *pending*, the immediate step to *agreed* is ruled out. If no other site is going to send a message, u is enforced to initiate a new proposal (action $a(u)$). Messages (v, u) then may provoke reactions of a site v, i.e., new proposals sent to its neighbors $g(v)$ (action $a(v)$). Then $g(u, v)$ may turn u to *quiet*. Again, this algorithm is terminated if and only if all sites are *agreed*.

35.3   A further variant of a consensus algorithm

In $\Sigma_{35.2}$, initiation of a new proposal, viz. occurrence of action $a(x)$, requires y to be *pending*. This side condition is replaced in $\Sigma_{35.3}$ by x to be *demanded*. Hence in $\Sigma_{35.2}$ a site x may initiate a new proposal even if x is *quiet*, whereas in $\Sigma_{35.3}$, x may initiate a new proposal even if x is *agreed*. Again, this algorithm is terminated if and only if all sites are agreed.

36   Phase Synchronization on Undirected Trees

36.1   The problem of phase synchronization

Network algorithms work frequently in *rounds* or *phases*: Each site eventually returns to its initial state, thus entering its next phase.

A synchronization mechanism is occasionally required, that guarantees synchronized execution of rounds: No site begins its $(k+1)$st phase unless all sites have completed their k-th phase. Stated differently, two sites that are busy at the same time are executing the same round. The crosstalk algorithm (Sect. 12) and all derived algorithms such as crosstalk based mutual exclusion (Sect. 13) and the distributed rearrangement algorithm (Sect. 25) have been examples for synchronized execution of rounds of two neighbored sites.

A phase synchronization algorithm is derived in the sequel, apt for any undirected tree network (such networks have been considered in Sect. 34.1 already).

36.2   The algorithm

Figure 36.1 provides phase synchronization on undirected trees: Each site alternates between two states, *busy* and *pending*. Initially, each site is busy in its zero round. A site u may communicate its actual round number n to one of its neighbors v by help of the *message* (v, u). A site u that is busy in its n-th round goes *pending* upon receiving messages from all but one neighbor, v, by action $a(u, v, n)$ (with the ordered set (x, y, i) of variables). As no message is available initially, the leaves of the underlying undirected tree start the algorithm. A *pending* site u at round n goes *busy* in round $n+1$ upon receiving the missing message (u, v), by action $b(u, v, n)$. Intuitively formulated, the leafs start waves of messages that are sent to inner nodes, thus involving more and more nodes. Eventually, two neighbored sites get messages from *all* their neighbors. In this case, the messages are *reflected*, i.e., are returned to their respective senders, which coincidently start their new round (transition b).

As an interesting observation, two identical messages may occur and must be treated as two different tokens (similar situations occurred in the alternating bit protocol of Sect. 27 and the sliding window protocol of Sect. 28). An example was the network formed

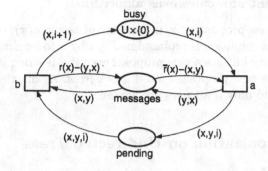

Figure 36.1. Phase synchronization

The following data accompanies the figure:

<u>sort</u> site	$W = W^{-1}$
<u>sort</u> message = site × site	$x,y \in U \to x\,W^*y$
<u>const</u> U: set of sites	$W_1 = U$
<u>const</u> W : set of (sites × sites)	$x_0 W\, x_1\, ...\, x_n W\, x_{n+1} \wedge$
<u>fct</u> r, $\bar{r}$: site → set of messages	$x_{i-1} \neq x_{i+1}$ for i=1,...,n
<u>var</u> x, y : site	$\to x_0 \neq x_n$
<u>var</u> i : nat	$r(x) = W(x) \times \{x\}$
	$\bar{r}(x) = r(x)^{-1}$

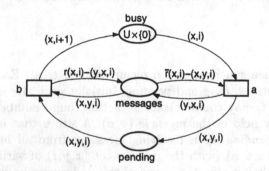

<u>sort</u> site	$W = W^{-1}$
<u>sort</u> message = site × site × nat	$x,y \in U \to x\,W^*y$
<u>const</u> U: set of sites	$W_1 = U$
<u>const</u> W : set of (sites × sites)	$x_0 W\, x_1\, ...\, x_n W\, x_{n+1} \wedge$
<u>fct</u> r, $\bar{r}$: site × nat → set of messages	$x_{i-1} \neq x_{i+1}$ for i=1,...,n
<u>var</u> x, y : site	$\to x_0 \neq x_n$
<u>var</u> i : nat	$r(x,i) = W(x) \times \{x\} \times \{i\}$
	$\bar{r}(x,i) = \{x\} \times W(x) \times \{i\}$

Figure 36.2. Messages with round number

$$A \text{——} B \text{——} C \qquad (1)$$

with a state consisting of the empty place *busy*, two copies of *messages* (B, C), and *pending* containing $(A, 0)$, $(B, 0)$, and $(C, 1)$. From the initial state, this state is reachable by the sequence of actions $a(A, B, 0)$, $a(B, C, 0)$, $a(C, B, 0)$, $b(C, B, 0)$, $a(C, B, 1)$. The corresponding state of $\Sigma_{36.1}$ includes two different *messages*, $(B, C, 0)$, and $(B, C, 1)$.

36.3 Variants of the algorithm

Occurrence of two identical messages in $\Sigma_{36.1}$ can be avoided: Just extend each message by the corresponding round number, as in $\Sigma_{36.2}$. The two messages (B, C) shown above to occur in the network (1) are $(B, C, 0)$ and $(B, C, 1)$ in $\Sigma_{36.2}$.

As a further variant, a pending site (u, v, n) of $\Sigma_{36.1}$ is not required to retain v; hence the version of $\Sigma_{36.3}$.

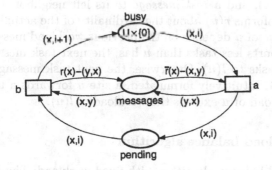

sort site	$W = W^{-1}$
sort message = site × site × nat	$x, y \in U \to x\,W^{*}\,y$
const U: set of sites	$W_1 = U$
const W: set of (sites × sites)	$x_0 W\, x_1 \dots x_n W\, x_{n+1} \wedge$
fct r, $\bar{r}$: site × nat → set of messages	$x_{i-1} \neq x_{i+1}$ for $i=1,\dots,n$
var x, y : site	$\to x_0 \neq x_n$
var i : nat	$r(x) = W(x) \times \{x\}$
	$\bar{r}(x) = r(x)^{-1}$

Figure 36.3. Pending without neighbor

37   Distributed Self Stabilization

37.1   Load balance in rings

A *service site* is intended to execute *tasks*, provided by the site's environment. At any reachable state a service site has its actual *workload*, i.e., a set of tasks still to be executed. The workload increases or decreases due to interaction with the environment.

Now assume a set of service sites, each one autonomously interacting with its environment. Their individual workload may be heavy or low in a given state, and it is worthwhile to *balance* them: A site with heavy workload may send some tasks to sites with less heavy workload. The overall workload in a set of service sites is *balanced* whenever the cardinality of the workload of two sites differs at most by one.

A distributed algorithm is constructed in the sequel, organizing load balancing in a set of service sites. The communication lines among sites are assumed to form a ring. Each agent u alternately sends a *workload message* to its right neighbor, $r(u)$, and a *task message* to its left neighbor, $l(u)$. A *workload message* of u informs $r(u)$ about the cardinality of the actual workload of u. A *task message* of u depends on the previous workload message of $l(u)$: If this message reports less tasks than u has, the next task message of u transfers one of u's tasks to $l(u)$. Otherwise, the next task message of u transfers no task to $l(u)$. Intuitively formulated, a site u forwards a task to $l(u)$ whenever the workload of u exceeds the workload of $l(u)$.

37.2   A distributed load balance algorithm

Figure 37.1 shows a load balance algorithm with fixed workload: The overall number of tasks remains constant. Each state of a site u is represented as a pair (u, n), with n the cardinality of u's actual workload. The task transfered from u to $l(u)$ by a task message $(l(u), 1)$, is not represented itself.

With the ordered set (x, i, j) of variables, action *inform right* describes communication with right neighbors: A site u with actually n tasks with action *informed right*(u, n, m) receives a task message (u, m) (with $m = 0$ or $m = 1$) from $r(u)$, updates its actual workload, n, and returns a corresponding workload message $(r(u), n + m)$ back to $r(u)$, indicating that u has actually $n + m$ tasks.

With the same ordered set of variables, actions *send left no task* and *send left one task* describe communication with left neighbors: A site u with actually n tasks receives a workload message (u, m) from $l(u)$, compares n and m, and returns a task with action *send left one task*(u, n, m) in case its actual workload, n, exceeds $l(u)$'s reported workload, m. Otherwise, u sends a task message with *send left no task*(u, n, m), to $l(u)$, containing no task.

Initially, each site u informs $r(u)$ about its actual workload.

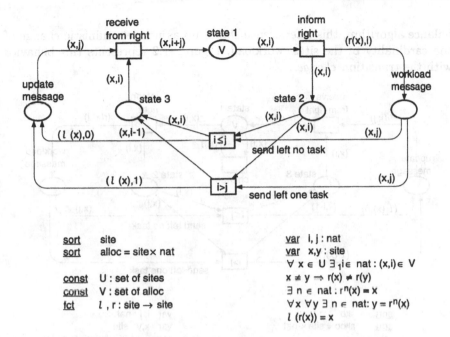

Figure 37.1. Distributed load balancing

37.3 Decisive properties of the algorithm

The above algorithm never terminates; each run is infinite. The overall work-load is eventually balanced, as described above. Two cases may be distinguished, depending on the overall workload $w := \Sigma_{v \in V_2} u$ and the number $|U|$ of sites:

In case w is a multiple of $|U|$, a state will be reached where transition *send left one task* remains inactive forever, and *state1*, *state2*, and *state3* together contain the tokens (u, n) with $u \in U$ and $n = \frac{w}{|U|}$. Otherwise a state will be reached where for all tokens (u, n) and (v, m) in *state1*, *state2*, and *state3* holds $|m - n| \leq 1$, and this remains valid forever. The algorithm behaves quite regularly: With initially V at *state1*, it evolves exactly one concurrent run. This run is strictly organized in rounds: All sites concurrently execute action *inform right* and produce a workload message for their respective right neighbor. Then all sites concurrently execute *send left no task* or *send left one task*, thus producing a task message for their respective left neighbor. Finally, *receive from right* completes a round.

37.4 Load balancing in a floating environment

The load balance algorithm should work concurrently to other parts of the service sites, in particular to increase and decrease of their respective workload. But it interferes with those actions. From the perspective of the load balance algorithm, this interference shines up as nondeterministic change of the cardinality of the sites' workload. Figure 37.2 represents this behavior with the transition *change*.

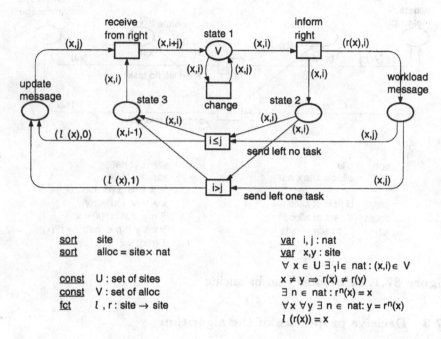

<u>sort</u>	site	<u>var</u>	$i, j : $ nat
<u>sort</u>	alloc $= $ site $\times$ nat	<u>var</u>	$x, y : $ site
		$\forall x \in U\ \exists_1 i \in$ nat $: (x,i) \in V$	
<u>const</u>	$U : $ set of sites	$x \neq y \Rightarrow r(x) \neq r(y)$	
<u>const</u>	$V : $ set of alloc	$\exists n \in$ nat $: r^n(x) = x$	
<u>fct</u>	$l , r : $ site $\rightarrow$ site	$\forall x\ \forall y\ \exists n \in$ nat $: y = r^n(x)$	
		$l (r(x)) = x$	

Figure 37.2. Distributed load balancing in a floating environment

The properties discussed in Sect. 37.3 are not guaranteed any more in $\Sigma_{37.2}$. A balanced state will be reached whenever *change* occurs seldomly, and do not drastically change the workload.

Part C
Analysis of Elementary System Models

The term "analysis" refers to means to show that a system has particular properties. Examples of systems and typical corresponding properties include

1. a mutual exclusion algorithm, preventing two sites coinciding in their respective critical sections;
2. a producer/consumer system, ensuring that the buffer never carries more than one item;
3. a pushdown device, guaranteeing the equation $pop(push(x)) = x$;
4. a mutual exclusion algorithm, ensuring that each pending site will eventually go critical;
5. a resource allocation procedure, eventually serving all demands;
6. a termination detection algorithm, establishing that all components of a distributed program are terminated.

System properties can be classified by various aspects. Two classes of properties will be considered in the sequel, called *state* and *progress* properties, respectively. Intuitively formulated, a state property stipulates that "something bad" never happens. A progress property stipulates that eventually "something good" will happen.

A slightly more formal explanation of safety and progress properties is based on global states $s \subseteq P_\Sigma$, be they reachable or not. For a set M of global states, call s an M-state iff $s \in M$. A *state property* has the typical form "each reachable state is an M-state". A *progress property* has the typical form "some M-state will eventually be reached" or, in its conditional form, "from each L-state some M-state will eventually be reached". The properties 1, 2, and 3 described above are state properties, whereas 4, 5, and 6 are progress properties.

Part C provides techniques to *describe* and to *prove* such properties. Particular *formulas* will be employed for this purpose, adopting concepts of temporal logic. However, we do not propose a full-fledged logic with completeness results, minimal sets of operators, or efficient model checking procedures. Formulas are just used to make intuitive statements and conclusions transparent and precise, this way deepening the reader's insight into the functioning of systems.

Technically, elementary properties (viz. valid formulas) will be derived from the static structure of a net, i.e., without constructing its runs. More involved valid formulas are gained by means of rules from already derived formulas.

VII. State Properties of Elementary System Nets

Here we consider properties of elementary system models that can be described by "at each reachable state holds p", with p a propositional expression. A typical example was mutual exclusion, with $p = \neg(crit_l \wedge crit_r)$. Techniques to verify such properties include *place invariants* and *initialized traps*.

38 Propositional State Properties

A property p that at each state a of an es-net Σ either *holds* or *fails* is called a *state property* of Σ. p will be said to *hold in* Σ iff p holds at each reachable state of Σ. Such properties can conveniently be represented by means of propositional formulas built from the conventional propositional operators $\neg$ (not), $\wedge$ (and), $\vee$ (or), $\rightarrow$ (implies), etc. Local states of es-nets will serve as *atoms* of such formulas.

38.1 Definition. *Let P be a set of symbols. Then the set $\mathrm{sf}(P)$ of state formulas over P is the smallest set of symbol sequences such that*

  i. $P \subseteq \mathrm{sf}(P)$, and
  ii. if $p, q \in \mathrm{sf}(P)$ then $\neg p \in \mathrm{sf}(P)$ and $(p \wedge q) \in \mathrm{sf}(P)$.

The conventional propositional shorthands will be employed, and sets of formulas are quantified as usual.

38.2 Notation. *Let P be a set of symbols and let $p, q \in \mathrm{sf}(P)$.*

  i. We usually write
     $(p \vee q)$ for $\neg(\neg p \wedge \neg q)$, and $(p \rightarrow q)$ for $(\neg p \vee q)$.
  ii. For a set $Q = \{q_1, \ldots q_n\} \subseteq \mathrm{sf}(P)$ of formulas we often write $\bigvee Q$ instead of $q_1 \vee \ldots \vee q_n$. Likewise, for $q_1 \wedge \ldots \wedge q_n$ we often write $\bigwedge Q$ or Q or $q_1 \ldots q_n$. The operator $\wedge$ is assumed to bind more strongly than any other operator.
  iii. p is an atom iff $p \in P$.

State formulas will be constructed from the local states of es-nets Σ. A state formula p will be said to *hold* or to *fail* in any global state a of Σ. Holding of state formulas is defined as follows:

38.3 Definition. *Let P be a set of symbols, let $a \subseteq P$ and let $p, q \in$ sf(P). Then $a \models p$ (p holds at a, a is a p-state) is inductively defined as follows:*

i. $a \models p$ *iff* $p \in a$, *for atoms* $p \in P$,
ii. $a \models \neg p$ *iff not* $a \models p$,
iii. $a \models p \wedge q$ *iff* $a \models p$ *and* $a \models q$,

This definition in fact returns the expected meaning for the shorthands of Notation 38.2, i.e.,

$$a \models p \vee q \text{ iff } a \models p \text{ or } a \models q,$$

and $a \models \bigwedge a$. $a \models p \rightarrow q$ iff $a \models q$ whenever $a \models p$

A formula p is said to hold in an es-net Σ iff p holds in each reachable state of Σ:

38.4 Definition. *Let Σ be an es-net and let $p \in$ sf(P_Σ). Then p is said to hold in Σ (p is a valid state property of Σ), written $\Sigma \models p$, iff $a \models p$ for each reachable state a of Σ.*

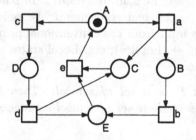

Figure 38.1. $\Sigma \models (B \rightarrow C) \wedge (A \rightarrow \neg C)$

As an example, in $\Sigma_{38.1}$ the two formulas $B \rightarrow C$ and $A \rightarrow \neg C$ hold. Further examples are $\Sigma_{9.1} \models A \vee B$, $\Sigma_{9.3} \models FC \rightarrow K$ and $\Sigma_{13.2} \models \neg(critical_l \wedge critical_r)$.

38.5 Lemma. *Let Σ be an es-net and let $p, q, r, s \in$ sf(P_Σ).*

i. $\Sigma \models p$ *and* $\Sigma \models q$ *iff* $\Sigma \models p \wedge q$
ii. *If* $\Sigma \models p$ *and* $\Sigma \models p \rightarrow q$ *then* $\Sigma \models q$
iii. *If* $\Sigma \models p \rightarrow q$ *and* $\Sigma \models r \rightarrow s$ *then* $\Sigma \models (p \wedge r) \rightarrow (q \wedge s)$.

Proof of this lemma is left as an exercise for the reader.

Some general properties of es-nets can be characterized by means of state formulas. Referring to Sect. 3, contact freeness is represented as follows:

38.6 Lemma. *An es-net Σ has no reachable contact state iff*

$$\Sigma \models \bigwedge_{t \in T}(\bigwedge {}^\bullet t \to \neg \bigvee(t^\bullet \setminus {}^\bullet t)).$$

Proof. Σ has no reachable contact state iff, for each reachable state a and each $t \in T_\Sigma$, if ${}^\bullet t \subseteq a$ then $(t^\bullet \setminus {}^\bullet t) \cap a = \emptyset$. This holds iff $\Sigma \models \bigwedge_{t \in T}(\bigwedge {}^\bullet t \to \neg \bigvee(t^\bullet \setminus {}^\bullet t))$. $\qquad\square$

A further formula describes that each reachable state enables at least one action:

38.7 Definition. *An es-net Σ is* stuck-free *iff* $\Sigma \models \bigvee_{t \in T}(\bigwedge {}^\bullet t \wedge \neg \bigvee(t^\bullet \setminus {}^\bullet t))$.

39   Net Equations and Net Inequalities

State properties can frequently be proven by means of equations and inequalities, which in turn can be derived from the static structure of any given es-net. To this end, each place p of an es-net Σ is taken as a *variable*, ranging over $\{0,1\}$, and each state $a \subseteq P_\Sigma$ is represented by its *characteristic function* $\underline{a} : P_\Sigma \to \{0,1\}$, with $a(p) = 1$ iff $p \in a$. Equations and inequalities with the form

$$n_1 \cdot p_1 + \ldots + n_k \cdot p_k = m \quad \text{and} \tag{1}$$

$$n_1 \cdot p_1 + \ldots + n_k \cdot p_k \geq m \tag{2}$$

will be constructed (where $p_1, \ldots, p_k$ are variables corresponding to P_Σ and $n_1, \ldots, n_k, m$ are integers), which holds in Σ if the characteristic function of each reachable state of Σ solves (1) and (2). Valid state properties can then be "picked up" from valid equations and inequalities.

39.1 Definition. *Let Σ be an es-net.*

i. *For each state $a \subseteq P_\Sigma$, the* characteristic function $\underline{a} : P_\Sigma \to \{0,1\}$ *of a is defined by $\underline{a}(p) = 1$ if $p \in a$ and $\underline{a}(p) = 0$ if $p \notin a$.*

ii. *Let $\{p_1, \ldots, p_k\} \subseteq P_\Sigma$ and let $p_1, \ldots, p_k$ be variables, ranging over $\{0,1\}$. Furthermore let $n_1, \ldots, n_k, m \in \mathbb{Z}$. Then*
$$\epsilon : \; n_1 \cdot p_1 + \ldots + n_k \cdot p_k = m$$
is a Σ-equation, and
$$\delta : \; n_1 \cdot p_1 + \ldots + n_k \cdot p_k \geq m$$
is a Σ-inequality.

iii. *Let ϵ and δ be as above, and let $a \subseteq P_\Sigma$ be a state. Then a solves ϵ iff $n_1 \cdot \underline{a}(p_1) + \ldots + n_k \cdot \underline{a}(p_k) = m$, and a solves δ iff $n_1 \cdot \underline{a}(p_1) + \ldots + n_k \cdot \underline{a}(p_k) \geq m$.*

iv. *A Σ-equation ϵ or a Σ-inequality δ is* valid *in Σ (ϵ or δ holds in Σ) iff each reachable state of Σ solves ϵ or δ, respectively.*

Addition and subtraction of valid Σ-equations and Σ-inequalities obviously retains validity.

We employ the usual conventions of integer terms, such as skipping $0 \cdot p_i$. For example, valid $\Sigma_{38.1}$-equations include $A + C + D = 1$, $B - C + E = 0$, and $2A + B + C + 2D + E = 2$. Likewise, the inequality $A + B + C + D + E \geq 1$ holds in $\Sigma_{38.1}$.

A valid Σ-equation immediately implies valid state properties. For example, the valid $\Sigma_{38.1}$-inequalities $A + C \leq 1$ and $B - C + E = 0$ yield $\Sigma_{38.1} \models A \rightarrow \neg C$ and $\Sigma_{38.1} \models B \rightarrow C$, respectively. Each valid Σ-equation ϵ implies a *strongest* valid state property of Σ, called the *state property of* ϵ. Most applications involve special cases of Σ-equations, with quite intuitive state properties. We start with the most general case, which may be skipped upon first reading, and will formulate the practically relevant cases as corollaries.

39.2 Definition. *Let Σ be an es-net and let ϵ be a Σ-equation.*

i. *Let the summands of ϵ be ordered such that ϵ reads $n_1 \cdot p_1 + \ldots + n_k \cdot p_k = m$, with $n_1 + \ldots + n_l = m$, for some $1 \leq l \leq k$. Then $p_1 \wedge \cdots \wedge p_l \wedge \neg p_{l+1} \wedge \cdots \wedge \neg p_k$ is a* standard formula *of ϵ.*

ii. *Let Π be the set of all standard formulas of ϵ. Then $\chi(\epsilon) := \bigvee \Pi$ is the* state property *of ϵ.*

For example, the equation $2A + B + C = 1$ has two standard formulas (up to propositional equivalence), $B \wedge \neg A \wedge \neg C$ and $C \wedge \neg A \wedge \neg B$. Likewise, $2A + B + C = 2$ has the standard formulas $A \wedge \neg B \wedge \neg C$ and $B \wedge C \wedge \neg A$, and the equation $A - B - C = 0$ has the standard formulas $A \wedge B \wedge \neg C$, $A \wedge C \wedge \neg B$, and $\neg A \wedge \neg B \wedge \neg C$.

The state property of each valid Σ-equation holds in Σ:

39.3 Theorem. *Let Σ be an es-net and let ϵ be a valid Σ-equation. Then $\Sigma \models \chi(\epsilon)$.*

Proof. i. Let $a \subseteq P_\Sigma$ be a reachable state of Σ. Then ϵ can be written

$$\epsilon : n_1 \cdot p_1 + \ldots + n_l \cdot p_l + n_{l+1} \cdot p_{l+1} + \ldots + n_k \cdot p_k = m, \text{ with}$$

$$p_1, \ldots, p_l \in a \text{ and } p_{l+1}, \ldots, p_k \notin a. \tag{1}$$

Then $m = n_1 \cdot \underline{a}(p_1) + \ldots + n_k \cdot \underline{a}(p_k)$ (by Def. 39.1(iii))
$= n_1 \cdot 1 + \ldots + n_l \cdot 1 + n_{l+1} \cdot 0 + \ldots + n_k \cdot 0$ (by def. of $\underline{a}$ and (1))
$= n_1 + \ldots + n_l$.

Then $\pi_a := p_1 \wedge \ldots \wedge p_l \wedge \neg p_{l+1} \wedge \ldots \wedge \neg p_k$ is a standard formula of ϵ. Furthermore,

$$a \models \pi_a \quad \text{(by (1) and Def. 38.3).} \tag{2}$$

ii. The set Π of all standard formulas contains the formula π_a for each reachable state a of Σ (by construction of π_a). Hence $\chi(\epsilon)$ holds for each reachable state (by (2)), which implies the Theorem (by Def. 38.4). $\square$

Applications mostly require propositional implications of state properties of quite special Σ-equations. The two most important cases are covered by the following corollary:

39.4 Corollary. *Let Σ be an es-net.*

i. *Let $p_1 + \ldots + p_k = 1$ be a valid Σ-equation. Then $\Sigma \models p_1 \vee \ldots \vee p_k$ and $\Sigma \models p_1 \rightarrow (\neg p_2 \wedge \ldots \wedge \neg p_k)$.*

ii. *With $n_1, \ldots, n_k > 0$, let $n_1 \cdot p_1 + \ldots + n_l \cdot p_l - n_{l+1} \cdot p_{l+1} - \ldots - n_k \cdot p_k = 0$ be a valid Σ-equation. Then $\Sigma \models (p_1 \vee \ldots \vee p_l) \rightarrow (p_{l+1} \vee \ldots \vee p_k)$.*

Proof.   i. The standard formulas of the given equation are $p_i \wedge \bigwedge_{j \neq i} \neg p_j$, for $i = 1, \ldots, k$. The properties claimed are implied by the disjunction of those formulas.

ii. For each standard formula $q_1 \wedge \cdots \wedge q_m \wedge \neg q_{m+1} \wedge \cdots \wedge \neg q_n$ of the given equation holds: If for some $1 \leq i \leq l$, $p_i \in \{q_1, \ldots, q_m\}$, then for some $m + 1 \leq j \leq n$, $p_j \in \{q_1, \ldots, q_m\}$, by construction of the equation. The property claimed is implied by the disjunction of those formulas. $\square$

By analogy to Def. 39.2, Theorem 39.3, and Corollary 39.4, each Σ-inequality δ can be assigned a set of standard formulas which yield a state property $\chi(\delta)$ that holds in Σ, provided δ is valid in Σ. Again, there is a most important special case:

39.5 Definition. *Let Σ be an es-net and let δ be a Σ-inequality.*

i. *Let the summands of δ be ordered such that δ reads $n_1 \cdot p_1 + \ldots + n_k \cdot p_k \geq m$, with $n_1 + \ldots + n_l \geq m$ for some $1 \leq l \leq k$. Then $p_1 \wedge \ldots \wedge p_l \wedge \neg p_{l+1} \wedge \ldots \wedge \neg p_k$ is a standard formula of δ.*

ii. *Let Π be the set of all standard formulas of δ. Then $\chi(\delta) := \bigvee \Pi$ is the state property of δ.*

For example, $2A + B - C \geq 2$ has the three standard formulas $A \wedge B \wedge C$, $A \wedge B \wedge \neg C$ and $A \wedge \neg B \wedge \neg C$.

The state property of each valid Σ-inequality holds in Σ:

39.6 Theorem. *Let Σ be an es-net and let δ be a valid Σ-inequality. Then $\Sigma \models \chi(\delta)$.*

Proof of this theorem tightly follows the proof of Theorem 39.3 and is left as an exercise for the reader.

The most important special case of the above theorem is captured by the following corollary:

39.7 Corollary. *Let Σ be an es-net and let $p_1 + \ldots + p_k \geq 1$ be a valid Σ-inequality. Then $\Sigma \models p_1 \vee \ldots \vee p_k$.*

40   Place Invariants of es-nets

Valid Σ-equations can be gained from solutions of systems of linear, homogeneous equations. To this end, a matrix $\underline{\Sigma}$ is assigned to each es-net Σ. This matrix employs the places and transitions of Σ as line and row indices:

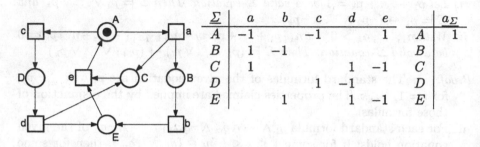

$\underline{\Sigma}$	a	b	c	d	e
A	-1		-1		1
B	1	-1			
C	1			1	-1
D			1	-1	
E		1		1	-1

	a_Σ
A	1
B	
C	
D	
E	

Figure 40.1. $\Sigma_{38.1}$ with matrix $\underline{\Sigma}$ and vector $\underline{a_\Sigma}$

40.1 Definition. *Let Σ be an es-net.*

i. *For $t \in T_\Sigma$ let $\underline{t}$ be the P_Σ-indexed vector where for each $p \in P_\Sigma$*

$$\underline{t}[p] := \begin{cases} +1 & \text{iff } p \in t^\bullet \setminus {}^\bullet t \\ -1 & \text{iff } p \in {}^\bullet t \setminus t^\bullet \\ 0 & \text{otherwise.} \end{cases}$$

ii. *Let $\underline{\Sigma}$ be the matrix with index sets P_Σ and T_Σ, such that for each $p \in P_\Sigma$ and $t \in T_\Sigma$, $\underline{\Sigma}[p,t] := \underline{t}[p]$.*

$\underline{\Sigma}$	a	b	c	d	e
A	-1		-1		1
B	1	-1			
C	1			1	-1
D			1	-1	
E		1		1	-1

	i_1	i_2	i_3	i_4
A	1	1	2	
B		1	1	1
C	1		1	-1
D	1	1	2	
E		1	1	1

Figure 40.2. Matrix and four place invariants of $\Sigma_{38.1}$

Figure 40.1 represents the matrix of $\Sigma_{38.1}$, as well as the characteristic function of the initial state. Intuitively, $\underline{\Sigma}[p,t]$ describes the change of the number of tokens on the place p upon any occurrence of t. The matrix $\underline{\Sigma}$ describes the static structure of Σ uniquely in case Σ is loop-free.

The matrix $\underline{\Sigma}$ of an es-net Σ will now be used to construct the system $\underline{\Sigma}^\tau \cdot x = 0$ of homogeneous linear equations. Here $\underline{\Sigma}^\tau$ denotes the transposed matrix of $\underline{\Sigma}$, 0 the P_Σ-indexed vector with zero entries, and $\cdot$ the usual inner product of matrices and vectors. The solutions of this system of equations are called *place invariants*.

40.2 Definition. *Let Σ be an es-net and let $i : P_\Sigma \rightarrow \mathbb{Z}$ be a mapping. Then i is called a* place invariant *of Σ iff $\underline{\Sigma}^\tau \cdot i = 0$ (i.e., for each $t \in T_\Sigma$, $\underline{t} \cdot i = \Sigma_{p \in P_\Sigma} \underline{t}[p] \cdot i(p) = 0$).*

Figure 40.2 recalls the matrix of $\Sigma_{40.1}$ and shows place invariants $i_1, \ldots i_4$ of $\Sigma_{40.1}$.

Each place invariant is now assigned its *characteristic equation*:

40.3 Definition. *Let Σ be an es-net with $P_\Sigma = \{p_1, \ldots, p_k\}$ and let i be a* place invariant *of Σ.*

   i. $m := i(p_1) \cdot \underline{a}_\Sigma(p_1) + \ldots + i(p_k) \cdot \underline{a}_\Sigma(p_k)$ is the initial value *of i.*
  ii. The Σ-equation $i(p_1) \cdot p_1 + \ldots + i(p_k) \cdot p_k = m$ is the equation *of i.*

In fact, the equation of a place invariant is valid in the underlying es-net Σ:

40.4 Theorem. *Let Σ be an es-net and let i be a place invariant of Σ. Then the equation of i is valid in Σ.*

Proof. Let $\epsilon : n_1 \cdot p_1 + \ldots + n_k \cdot p_k = m$ be the equation of i. Then for each $t \in T_\Sigma$,

$$\Sigma_{i=1}^k n_i \cdot \underline{t}[p_i] = 0, \tag{1}$$

by Defs. 40.3(ii) and 40.2.
Now let $a \xrightarrow{t} b$ be any step of Σ. Then for each $p \in P_\Sigma$,

$$\underline{b}(p) = \underline{a}(p) + \underline{t}[p] \tag{2}$$

by Def. 40.1(i). Furthermore,

   if a solves ϵ then b solves ϵ, too, $\tag{3}$

shown as follows. Assume a solves ϵ. Then

$$m = \Sigma_{i=1}^k n_i \cdot \underline{a}(p_i) \qquad\qquad \text{(by Def. 39.1(iii))}$$

$$= (\Sigma_{i=1}^k n_i \cdot \underline{a}(p_i)) + (\Sigma_{i=1}^k n_i \cdot \underline{t}[p_i]) \qquad\qquad \text{(by (1))}$$

$$= \Sigma_{i=1}^k (n_i \cdot \underline{a}(p_i)) + (n_i \cdot \underline{t}[p_i]) = \Sigma_{i=1}^k n_i \cdot (\underline{a}(p_i) + \underline{t}[p_i])$$

$$= \Sigma_{i=1}^k n_i \cdot \underline{b}(p_i) \qquad\qquad \text{(by (2))}.$$

Then b solves ϵ $\qquad\qquad$ (by Def. 39.1(iii)).

To show the theorem, let a be any reachable state of Σ. Then there exists an interleaved run $a_0 \xrightarrow{t_1} a_1 \xrightarrow{t_2} \ldots \xrightarrow{t_l} a_l$ of Σ with $a_0 = a_\Sigma$ and $a_l = a$. Then a_0 solves ϵ (by Def. 40.3(i)). Then each $\underline{a}_j$ $(j = 0, \ldots, l)$ solves ϵ (by (3) and induction on j). Hence the theorem, by Def. 39.1(iv). $\qquad\qquad\square$

The equations of $i_1, \ldots, i_4$ as given in Fig. 40.2 are $A + C + D = 1$, $A + B + D + E = 1$, $2A + B + C + 2D + E = 2$, and $B - C + E = 0$, respectively. They are in fact valid, due to Theorem 40.4.

Important state properties can frequently be proven by means of equations of place invariants with entries ranging over $\{0, 1\}$ and initial value equal to 1. Examples are i_1 and i_2 of Fig. 40.2. The equation of such a place invariant is formed $p_1 + \ldots + p_k = 1$, with $P = \{p_1, \ldots, p_k\} \subseteq P_\Sigma$. It can be graphically depicted with boldfaced arcs adjacent to places in P. Figure 40.3 shows examples. The places in P together with the adjacent transitions in $P^\bullet \cup {}^\bullet P$ from a "subnet" with $|{}^\bullet t| = |t^\bullet| = 1$ for each action t. Occurrence of t (which may depend on local states not in P) then swaps the unique token within P.

According to Corollary 39.4(i), such place invariants yield valid state properties. For example, with invariant i_1 of Fig. 40.2, the state formulas $A \vee C \vee D$ and $A \rightarrow (\neg C \wedge \neg D)$ hold in $\Sigma_{40.1}$.

Slightly more generally, many place invariants have entries ranging over $\{-1, 0, +1\}$, with initial value 0. In Fig. 40.2, i_4 is an example. With Corollary 39.4(ii), such place invariants yield formulas of form $(p_1 \vee \ldots \vee p_l) \rightarrow (p_{l+1} \vee \ldots \vee p_k)$. For example, i_4 of Fig. 40.2 yields $(B \vee E) \rightarrow C$, and $-i_4$ implies $C \rightarrow (B \vee E)$.

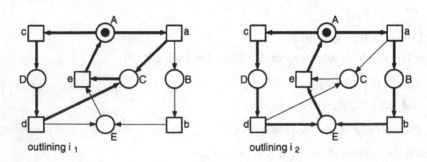

outlining i_1 outlining i_2

Figure 40.3. Outlining invariants i_1 and i_2 as given in Fig. 40.2

41   Some Small Case Studies

41.1   State properties of producer/consumer systems

The producer/consumer systems $\Sigma_{8.1}$, $\Sigma_{9.1}$, $\Sigma_{9.2}$, and $\Sigma_{9.3}$ consist essentially of small circles, synchronized along common actions. Each circle is characterized by a place invariant, in fact a characteristic vector. In $\Sigma_{9.1}$ their equations are

the producer's invariant i_1: $A + B = 1$ (1)

the first buffer cell's invariant i_2: $C + D = 1$ (2)

the second buffer cell's invariant i_3: $E + F = 1$ (3)

the consumer's invariant i_4: $G + H = 1$ (4)

Hence in each reachable state, the producer is either *ready to produce* or *ready to deliver*, each buffer cell is either *empty* or *filled*, and the consumer is either *ready to remove* or *ready to consume*.

It is not difficult to realize that the above four invariants also apply to $\Sigma_{9.2}$ and $\Sigma_{9.3}$.

The rest of this section is dedicated to the "optimality" of $\Sigma_{9.3}$, as explained in Sect. 9: The producer is never forced to access a filled buffer cell while the other cell is ready to deliver. Furthermore, if at least one buffer cell is empty, then the token on J or K "points at" an empty buffer cell. This is represented by

$$(B \wedge (E \vee C)) \to (BEJ \vee BCK). \tag{5}$$

Proof of (5) is based on three invariants (6)-(8), of which (7) is outlined in Fig. 41.1:

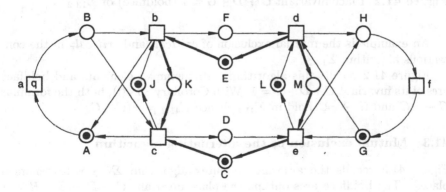

Figure 41.1. Place invariant $E + K - C - M = 0$ of $\Sigma_{9.3}$

$$J + K = 1, \text{ hence } \Sigma_{9.3} \models J \vee K. \tag{6}$$

$$E + K - C - M = 0, \text{ hence } \Sigma_{9.3} \models EK \rightarrow CK. \tag{7}$$

$$C + J - E - L = 0, \text{ hence } \Sigma_{9.3} \models CJ \rightarrow EJ. \tag{8}$$

Now we derive

$$\Sigma_{9.3} \models (E \vee C) \rightarrow (E \vee C) \wedge (J \vee K), \text{ by (6)}. \tag{9}$$

$$\Sigma_{9.3} \models (E \vee C) \rightarrow EJ \vee EK \vee CJ \vee CK, \text{ by (9)}. \tag{10}$$

$$\Sigma_{9.3} \models (E \vee C) \rightarrow EJ \vee CK, \text{ by (10),(7) and (8)}. \tag{11}$$

Now (5) follows from (11) by propositional logic.

41.2 Mutual exclusion of the contentious algorithms

Mutual exclusion of two local states p and q of some es-net Σ is apparently represented by

$$\Sigma \models \neg(p \wedge q). \tag{1}$$

This can frequently be proven by means of place invariants.

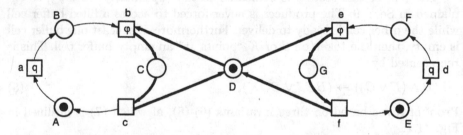

Figure 41.2. Place invariant $C + D + G = 1$ (boldface) of $\Sigma_{13.2}$

An example is the mutual exclusion of *critical$_l$* and *critical$_r$* in the contentious algorithm $\Sigma_{13.2}$.

Figure 41.2 recalls this algorithm, with renamed elements and boldface arcs of its invariant $C + D + G = 1$. With Corollary 39.4(i), both the formulas $C \rightarrow \neg G$ and $G \rightarrow \neg C$ hold in $\Sigma_{41.2}$, hence $\Sigma_{41.2} \models \neg(C \wedge G)$.

41.3 Mutual exclusion of the alternating algorithm

Figure 41.3 recalls the alternating mutex algorithm $\Sigma_{13.3}$ with renamed places. The boldface arcs outline the place invariant $C + D + E + H = 1$ which implies $\neg(C \wedge H)$, viz. mutual exclusion of *critical$_l$* and *critical$_r$* in $\Sigma_{13.3}$.

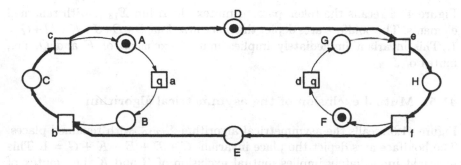

Figure 41.3. Place invariant (boldface) of the alternating mutex algorithm $\Sigma_{13.3}$

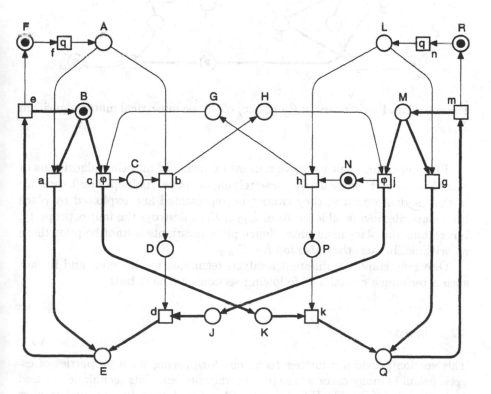

Figure 41.4. Place invariant (boldface) of the token passing mutex algorithm $\Sigma_{13.5}$

41.4   Mutual exclusion of the token passing algorithm

Figure 41.4 recalls the token passing mutex algorithm $\Sigma_{13.5}$, with renamed elements. The boldface arcs depict the place invariant $B+E+J+K+M+Q = 1$. This invariant immediately implies mutual exclusion of E and Q, i.e., mutex of $\Sigma_{13.5}$.

41.5   Mutual exclusion of the asymmetrical algorithm

Figure 41.5 recalls the asymmetrical algorithm $\Sigma_{13.10}$, with renamed places. The boldface arcs depict the place invariant $C + D + E + K + G = 1$. This invariant immediately implies mutual exclusion of C and K, i.e., mutex of $\Sigma_{13.10}$.

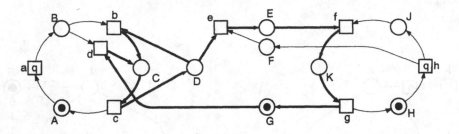

Figure 41.5. Place invariant (boldface) of the asymmetrical mutex algorithm $\Sigma_{13.9}$

Place invariants fail to prove mutual exclusion of all other algorithms of Sect. 13, however, particularly those relying on loops. As loops don't show up in the incidence matrix, they cannot be represented and exploited by place invariants. Removing all loops from $\Sigma_{13.4}$, $\Sigma_{13.8}$ destroys the mutex property, but retains the place invariants. Hence place invariants cannot help for those algorithms. In fact, they also fail for $\Sigma_{13.6}$.

One may employ a different analysis technique to this end, and in fact such a technique exists. The following section has the details.

42   Traps

This section provides a further technique for proving state properties of es-nets, useful in many cases where place invariants fail. This technique is based on *traps*, i.e., on subsets $P \subseteq P_\Sigma$ with $P^\bullet \subseteq {}^\bullet P$. A trap $\{p_1, \ldots, p_k\}$ implies the valid inequality $p_1 + \ldots + p_k \geq 1$ and hence the state formula $p_1 \vee \ldots \vee p_k$, provided one of its places belongs to the initial state. Hence we are mostly interested in *initialized* traps:

42.1 Definition. *Let Σ be an es-net and let $P \subseteq P_\Sigma$.*

  i. P is a trap *iff $P \neq \emptyset$ and $P^\bullet \subseteq {}^\bullet P$.*
 ii. P is initialized *iff $P \cap a_\Sigma \neq \emptyset$*

Figure 42.1 shows an example of an initialized trap, $\{A, D\}$. Figure 42.2 out-

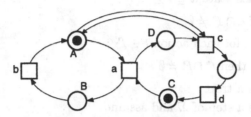

Figure 42.1. Initialized trap $\{A, D\}$

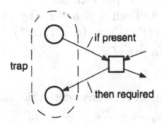

Figure 42.2. Graphical outline of a trap

lines the requirement for traps graphically. As a further example, $\{A, C, D, E\}$ is an initialized trap of $\Sigma_{40.1}$.
    Basic rules on sets imply:

42.2 Proposition. *The union of traps of a net N is again a trap of N.*

Proof. Let A and B be traps of N. Then $(A \cup B)^\bullet = A^\bullet \cup B^\bullet \subseteq {}^\bullet A \cup {}^\bullet B = {}^\bullet(A \cup B)$. $\qquad\qquad\square$

Each trap is now assigned *its inequality*, by analogy to the equation of place invariants, as defined in Def. 40.3.

42.3 Definition. *Let Σ be an es-net and let $P = \{p_1, \ldots, p_k\} \subseteq P_\Sigma$ be a trap of Σ. Then*

$$p_1 + \ldots + p_k \geq 1$$

is the inequality *of P.*

    In fact, the inequality of an initialized trap is valid:

42.4 Theorem. *Let Σ be an es-net and let $P \subseteq P_\Sigma$ be an initialized trap. Then the inequality of P is valid in Σ.*

Proof.   i. Let $\delta : p_1 + \ldots + p_k \geq 1$ be the inequality of P, and let $a \subseteq P_\Sigma$ be a state of Σ. Then a solves δ iff $\underline{a}(p_1) + \ldots + \underline{a}(p_k) \geq 1$ iff $\underline{a}(p_i) = 1$ for at least one $1 \leq i \leq k$ iff $p_i \in a$ for at least one $1 \leq i \leq a$ iff $P \cap a \neq \emptyset$. Hence for each state $a \subseteq P_\Sigma$,

$$a \text{ solves } \delta \text{ iff } a \cap P \neq \emptyset. \tag{1}$$

Furthermore, for each action $t \in T_\Sigma$:

$$\text{If } {}^\bullet t \cap P \neq \emptyset \text{ then } t^\bullet \cap P \neq \emptyset \tag{2}$$

because P is a trap.

   ii. Let $a \xrightarrow{t} b$ be a step of Σ and assume

$$a \cap P \neq \emptyset. \tag{3}$$

We distinguish two cases:
If ${}^\bullet t \cap P \neq \emptyset$ then $\emptyset \neq t^\bullet \cap P$ (by (2)) $\subseteq ((a \setminus {}^\bullet t) \cap P) \cup (t^\bullet \cap P) = ((a \setminus {}^\bullet t) \cup t^\bullet) \cap P = b \cap P$ (by Def. 3.1).
Otherwise ${}^\bullet t \cap P = \emptyset$. Then $\emptyset \neq a \cap P$ (by (3)) $= (a \cap P) \setminus ({}^\bullet t \cap P) \subseteq ((a \cap P) \setminus ({}^\bullet t \cap P)) \cup (t^\bullet \cap P) = ((a \setminus {}^\bullet t) \cup t^\bullet) \cap P = b \cap P$ (by Def. 3.1).
Hence for each step $a \xrightarrow{t} b$ of Σ,

$$\text{If } a \cap P \neq \emptyset \text{ then } b \cap P \neq \emptyset. \tag{4}$$

   iii. To show the theorem, let a be any reachable state of Σ. Then there exists an interleaved run $a_0 \xrightarrow{t_1} a_1 \xrightarrow{t_2} \ldots \xrightarrow{t_l} a_l$ of Σ with $a_0 = a_\Sigma$ and $a_l = a$. Then $a_0 \cap P \neq \emptyset$ (by the theorem's assumption and Def. 42.1). Then $a_i \cap P \neq \emptyset$ for $i = 1, \ldots, l$ by (4) and induction on i. Then a_l solves δ (by (i)), hence a solves δ.   □

The inequality of a trap can be combined with inequalities of other traps and with equations of place invariants. A small but typical example is the proof of $\Sigma_{42.1} \models B \to D$, hence the equation

$$B \leq D \tag{1}$$

must hold. Figure 42.1 yields the place invariant

$$A + B = 1. \tag{2}$$

Furthermore, the initialized trap $\{A, D\}$ yields the inequality

$$A + D \geq 1. \tag{3}$$

Subtracting (3) from (2) then yields

$$B - D \leq 0 \tag{(2) - (3)}$$

which immediately implies (1).

The inequality of an initialized trap P yields the formula $\bigvee P$, according to Corollary 39.7.

42.5 Corollary. *Let Σ be an es-net and let $P \subseteq P_\Sigma$ be an initialized trap. Then $\Sigma \models \bigvee P$.*

Proof. The proposition follows from Def. 42.3 and Corollary 39.7. $\square$

43 Case Study: Mutex

Mutual exclusion has been shown for a number of algorithms in Sect. 41. Here we show mutual exclusion for the remaining algorithms of Sect. 13 by combining equations and inequalities of place invariants and traps.

43.1 Mutex of the state testing algorithm

Figure 43.1 redraws the state testing algorithm, renaming its local states and outlining its trap $\{D, E\}$ by boldfaced arcs. Mutual exclusion in this representation reads $\Sigma_{43\,1} \models \neg(C \wedge H)$, hence the equation

$$C + H \leq 1 \tag{1}$$

must hold. Proof of (1) is based on two elementary place invariants with the equations

$$C + D = 1, \tag{2}$$

$$E + H = 1, \tag{3}$$

and the initialized trap $\{D, E\}$ outlined in Fig. 43.1 which yields

$$D + E \geq 1. \tag{4}$$

Subtracting the inequality (4) from the sum of the equations (2) and (3) yields

$$C + D + E + H - D - E \leq 1 + 1 - 1 \qquad (2) + (3) - (4)$$

which immediately reduces to (1).

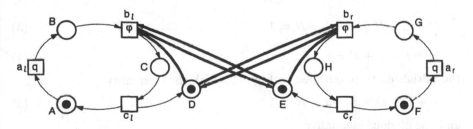

Figure 43.1. Trap $\{D, E\}$ of the state testing *mutex* algorithm $\Sigma_{12.4}$

43.2   Mutex of the round-based algorithm

Figure 43.2 redraws the round-based mutex algorithm $\Sigma_{13.6}$, renaming its local states and outlining one of its traps by boldfaced arcs. Mutual exclusion

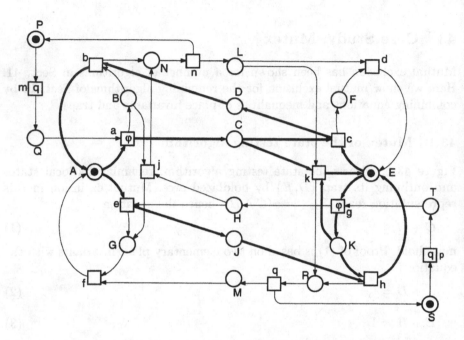

Figure 43.2. Trap $\{A, C, D, E, J, K\}$ of the round-based *mutex* algorithm $\Sigma_{13.6}$

in this representation reads $\Sigma_{43.2} \models \neg(N \wedge R)$, hence the equation

$$N + R \leq 1 \tag{1}$$

must hold. Proof of (1) is based on three place invariants with the equations

$$A + C + D + J + M + R = 1, \tag{2}$$

$$D + E + H + J + L + N = 1, \tag{3}$$

$$K + R + S + T = 1, \tag{4}$$

the initialized trap, outlined in Fig. 43.2 with the inequality

$$A + C + D + E + J + K \geq 1, \tag{5}$$

and the obvious inequality

$$D + H + J + L + M + S + T \geq 0. \tag{6}$$

Thus we obtain by (2) + (3) + (4) − (5) − (6)

$N + 2R \leq 2$, i.e., $\frac{N}{2} + R \leq 1$. $\hspace{6cm}$ (7)

(7) is equivalent to (1), because N and R vary over $\{0, 1\}$.

43.3 Mutex of Peterson's algorithm

Figure 43.3 redraws Peterson's mutex algorithm $\Sigma_{13.7}$, renaming its local states and outlining its trap $\{C, F, G, M\}$ by boldfaced arcs. Mutual exclusion

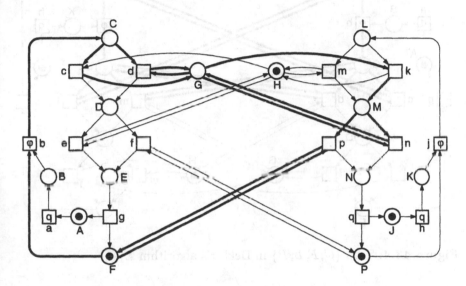

Figure 43.3. Trap $\{C, F, G, M\}$ in Peterson's algorithm $\Sigma_{13.7}$

in this representation reads $\Sigma_{43.3} \models \neg(E \wedge N)$, hence the equation

$E + N \leq 1$ $\hspace{7cm}$ (1)

must hold. Proof of (1) is based on three elementary place invariants with the equations

$G + H = 1,$ $\hspace{7cm}$ (2)

$C + D + E + F = 1,$ $\hspace{6cm}$ (3)

$L + M + N + P = 1,$ $\hspace{6cm}$ (4)

and two initialized traps, one of which is outlined in Fig. 43.3, which yield

$C + F + G + M \geq 1,$ $\hspace{6cm}$ (5)

$L + P + H + D \geq 1.$ $\hspace{6cm}$ (6)

Subtraction of (5) and (6) from the place invariants' sum, i.e., $(2) + (3) + (4) - (5) - (6)$, then immediately yields (1).

43.4   Mutex of Dekker's algorithm

Figure 43.4 renames the local states of Dekker's mutex algorithm $\Sigma_{13.8}$ and outlines its trap $\{F, C, P, L\}$. Mutual exclusion then reads $\Sigma_{43.4} \models \neg(D \wedge M)$,

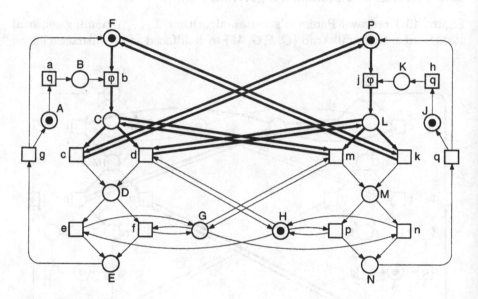

Figure 43.4. Trap $\{C, F, L, P\}$ in Dekker's algorithm $\Sigma_{13.8}$

hence the equation

$$D + M \leq 1 \tag{1}$$

must hold. Proof of (1) is based on two elementary place invariants with the equations

$$F + C + D + E = 1, \tag{2}$$

$$P + L + M + N = 1, \tag{3}$$

the initialized trap, outlined in Fig. 43.3, with the inequality

$$F + C + P + L \geq 1, \tag{4}$$

and the obvious inequality

$$E + N \geq 0. \tag{5}$$

Then (1) follows with (2) + (3) − (4) − (5).

43.5   Mutex of Owicki/Lamport's algorithm

Figure 43.5 renames the local states of Owicki/Lamport's algorithm $\Sigma_{13.9}$ and outlines its trap $\{C, F, G, K\}$. Mutual exclusion of writing and reading then reads $\Sigma_{43.5} \models \neg(D \wedge L)$, hence the equation

$$D + L \le 1 \tag{1}$$

must hold. Proof of (1) is based on two elementary place invariants with the equations

$$C + D + F = 1, \tag{2}$$

$$G + K + L = 1, \tag{3}$$

and an initialized trap, outlined in Fig. 43.5, with the inequality

$$C + F + G + K \ge 1. \tag{4}$$

Subtraction of (4) from the sum of (2) and (3), i.e., (2) + (3) − (4), then reduces to (1).

This completes the proof of mutual exclusion for all algorithms of Sect. 13.

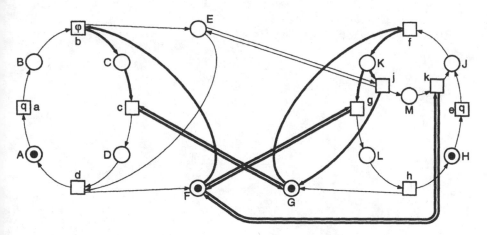

Figure 43.5. Trap $\{C, F, G, K\}$ in Owicki/Lamport's asymmetrical mutex algorithm $\Sigma_{13.9}$

VIII. Interleaved Progress of Elementary System Nets

As explained in the introductory text of Part C, a progress property of an es-net Σ stipulates for a given set M of states that one of them will eventually be reached. In its conditional form, a progress property stipulates that, starting at any state in some set L of states, a state in M will eventually be reached.

The notion of progress can be based on interleaved runs as well as on concurrent runs. This section sticks to the interleaved version. Concurrent variants will follow in Chap. IX.

44   Progress on Interleaved Runs

We consider progress properties that are constructed from two state properties, p and q: The progress property p *leads to* q (written $p \mapsto q$) holds in an interleaved run r of some es-net Σ iff each p-state of r is eventually followed by some q-state. Furthermore, $p \mapsto q$ is said to hold in Σ iff $p \mapsto q$ holds in each run of Σ. For example, the evolution property of a mutex algorithm Σ (cf. Sect. 13) then reads

$$\Sigma \models pending \mapsto critical \tag{1}$$

Technically, *leads-to* formulas are constructed from state formulas (cf. Def. 38.1):

44.1 Definition. *Let P be a set of symbols and let $p, q \in \mathrm{sf}(P)$ be state formulas over P. Then the symbol sequence $p \mapsto q$ ("p leads to q") is a leads-to formula over P. The set of all such formulas is denoted $lf(P)$.*

Leads-to formulas are interpreted over interleaved runs and over es-nets:

44.2 Definition. *Let Σ be an es-net and let $p \mapsto q \in lf(P_\Sigma)$ be a leads-to formula.*

  i. *$p \mapsto q$ is said to hold in an interleaved run w of Σ (written $w \models p \mapsto q$) iff to each p-state of w with index i there exists a q-state in w with some index $j \geq i$.*
  ii. *$p \mapsto q$ is said to hold in Σ (written $\Sigma \models p \mapsto q$) iff $w \models p \mapsto q$ for each interleaved run w of Σ.*

For example, in $\Sigma_{44.1}$, $AB \mapsto E$, $A \mapsto CD$, and $A \mapsto E$ hold, but not $AB \mapsto AD$. In $\Sigma_{44.2}$, $ABC \mapsto (F \vee G)$ and $AB \mapsto (F \vee DG)$ hold.

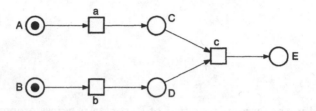

Figure 44.1. $A \mapsto E$

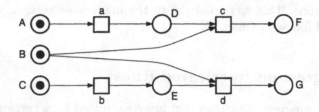

Figure 44.2. $ABC \mapsto (F \vee G)$, and $AB \mapsto (F \vee DG)$

Essential properties of the case studies of Chap. II can be formulated by means of *leads-to* formulas. Examples include:

- for buffer systems Σ as in Figs. 9.1–9.3: Each producer *ready to deliver* will eventually return to *ready to produce*: $\Sigma \models B \mapsto A$ (but $A \mapsto B$ does not hold there!);
- for actor/responder systems Σ as in Figs. 12.1, 12.2, 12.3, 12.5: The left agent will eventually return to its local state: $\Sigma \models \neg local_l \mapsto local_l$;
- for mutual exclusion algorithms Σ as in Figs. 13.4–13.10: Each pending site will eventually be critical: $\Sigma \models pending \mapsto critical$ or each prepared writer will eventually be writing.

The leads-to operator exhibits a couple of useful properties:

44.3 Lemma. *Let Σ an es-net and let $p, q, r \in \mathrm{sf}(P_\Sigma)$.*

i. *If $\Sigma \models p \to q$ then $\Sigma \models p \mapsto q$;*
ii. *$\Sigma \models p \mapsto p$;*
iii. *If $\Sigma \models p \mapsto q$ and $\Sigma \models q \mapsto r$ then $\Sigma \models p \mapsto r$;*
iv. *If $\Sigma \models p \mapsto r$ and $\Sigma \models q \mapsto r$ then $\Sigma \models (p \vee q) \mapsto r$.*

Brackets will be avoided in progress formulas by the assumption that the progress operator $\mapsto$ binds more weakly than any propositional operator. For example, $p \wedge q \mapsto r \vee s$ will stand for $(p \wedge q) \mapsto (r \vee s)$.

Proof of this lemma is left as an exercise for the reader.

45 The Interleaved Pick-up Rule

Section 44 introduced means to *represent* leads-to properties. Now we pursue a method to *prove* such properties. To this end we suggest a technique to "pick up" simple valid leads-to formulas from the static structure of a net. Further valid formulas can be derived from already established state- and progress properties by help of the next chapter's *proof graphs*.

The forthcoming technique is based on the (admittedly quite obvious) observation that either an enabled, progressing action occurs itself, or one of its attached neighbors occurs.

As an example, assume the following piece

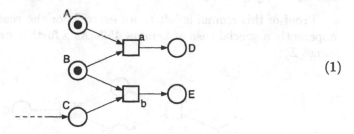

$$(1)$$

of an es-net Σ. $\{A, B\}$ enables the progressing action a; hence either a or b occurs eventually. Represented in the framework of Sect. 44, we gain $\Sigma \models AB \mapsto D \vee AE$. In general:

45.1 Lemma. *Let Σ be an es-net and let $t \in T_\Sigma$ be progressing. Then $\Sigma \models$* $^\bullet t \mapsto \bigvee_{u \in (^\bullet t)^\bullet} \mathrm{eff}(^\bullet t, u)$.

Proof of this lemma is left as an exercise for the reader.

More generally, we may start out with any *progress prone* set of local states. As an example, in

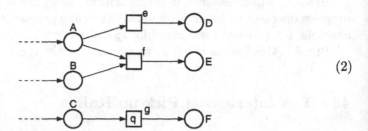

$$(2)$$

the state AC enables e, hence either of e, f, or g will eventually occur. Hence $\Sigma \models AC \mapsto (CD \vee CE \vee AF)$.

45.2 Definition. *Let Σ be an es-net and let $Q \subseteq P_\Sigma$. Then Q is progress prone iff Q enables at least one progressing action of Σ.*

As an example, AC is progress prone in (2) whereas BC is not.

45.3 Lemma. *Let Σ be an es-net and let $Q \subseteq P_\Sigma$ be progress prone. Then $\Sigma \models Q \mapsto \bigvee_{u \in Q^\bullet} \mathrm{eff}(Q, u)$.*

Proof of this lemma is left as an exercise for the reader. Lemma 45.1 is apparently a special case of Lemma 45.3. As a further example consider the es-net $\Sigma =$

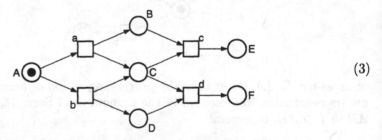

$$(3)$$

Def. 45.2 implies $\Sigma \models BC \mapsto E \vee BF$. The overall structure of Σ furthermore implies $\Sigma \models B \rightarrow \neg D$. Hence d is *prevented* in the state BC, and c is the only action to occur. Thus even $\Sigma \models BC \mapsto E$ holds. Generally, a set Q of local states of an es-net Σ *prevents* an action t iff t is not enabled at any reachable state a with $Q \subseteq a$. This holds true iff $\Sigma \models Q \rightarrow \neg^\bullet t$. A *change set* of Q then includes all $t \in Q^\bullet$ that are not prevented by Q:

45.4 Definition. *Σ be an es-net and let $Q \subseteq P_\Sigma$.*

i. Q prevents an action $t \in T_\Sigma$ iff $\Sigma \models Q \rightarrow \neg(^\bullet t)$.
ii. $U \subseteq T_\Sigma$ is a change set of Q iff $U \neq \emptyset$ and Q prevents each $t \in Q^\bullet \setminus U$.

$Q^\bullet$ is obviously a change set of Q. In the net Σ as given in (3), BC is progress prone, whereas BD is not. BC prevents d and CD prevents c. The set $\{c, d\}$ as well as the set $\{c\}$ are change sets of BC. The set $\{a, b\}$ is a change set of A whereas $\{a\}$ is no change set of A.

The following theorem describes the most general case to pick up leads-to formulas from the static structure of a net: Each change set of a progress prone set Q implies a leads-to formula.

45.5 Theorem. *Let Σ be an es-net, let $Q \subseteq P_\Sigma$ be progress prone and let U be a change set of Q in Σ. Then*

$$\Sigma \models Q \mapsto \bigvee_{u \in U} \mathrm{eff}(Q, u).$$

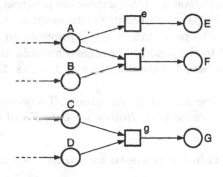

Figure 45.1. $AC \mapsto EC$ holds, provided $A \rightarrow \neg D$ and $C \rightarrow \neg B$ hold.

Proof. Let $w = a_0 \xrightarrow{t_1} a_1 \xrightarrow{t_2} \ldots$ be an interleaved run of Σ and let $a_i \models Q$ for some index i of w. Then there exists a Q-enabled progressing action, t.

- Then a_i enables t (because $^\bullet t \subseteq Q \subseteq a_i$).
- Then there exists an index $j > i$ with $t_j \in (^\bullet t)^\bullet$ (by Def. 8.2(i) and Def. 6.1).
- Then there exists an index $l \leq j$ with $t_l \in Q^\bullet$. Let k be the smallest such index.
- Then $a_j \models Q$ for all $i \leq j < k$, and particularly $a_{k-1} \models Q$.
- Then $t_k \in U$ (by Def. 45.4), and furthermore $a_k \models \mathrm{eff}(Q, t_k)$.
- Hence $a_k \models \bigvee_{u \in U} \mathrm{eff}(Q, u)$.

The theorem now follows from Def. 44.2. $\square$

Lemma 45.3 is apparently a special case of Theorem 45.5 (with $U = Q^\bullet$).

A further, slightly more involved example is shown in Fig. 45.1: Assuming $A \rightarrow \neg D$ and $C \rightarrow \neg B$ to hold in $\Sigma_{45.1}$, $\{A, C\}$ prevents both g and f. Hence $U = \{e\}$ is a change set of $\{A, C\}$. Furthermore, $\{A, C\}$ is progress prone, hence $\Sigma_{45.1} \models AC \mapsto EC$. As a final example, $\Sigma_{45.2} \models \neg B \vee \neg E$, hence $\{A\}$ prevents b, and $\Sigma_{45.2} \models A \mapsto C$ follows from Theorem 45.5.

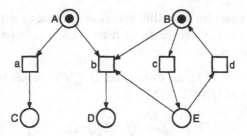

Figure 45.2. $\Sigma \models A \mapsto C$

Small change sets U generate more expressive progress formulas than large ones. However, it is occasionally useful not to insist on minimal change sets U: It may be difficult to prove that a set Q prevents an action $t \in Q^{\bullet}$, and the contribution of t to the generated progress formula may be irrelevant for the intended use. In fact, the special cases of Lemma 45.1 or Lemma 45.3 frequently suffice.

According to its use in correctness proofs, Theorem 45.5 suggests constructing a valid progress formula from a set $Q \subseteq P_{\Sigma}$ of places according to the following schema:

45.6 The pick-up rule for leads-to formulas. Let Σ be an es-net and let $Q \subseteq P_{\Sigma}$.

1. Make sure Q enables some progressing action $t \in T_{\Sigma}$ (i.e., $^{\bullet}t \subseteq Q$).
2. Starting with $U := Q^{\bullet}$, identify some actions prevented by Q and remove them from U.
3. With the remaining set $U \subseteq Q^{\bullet}$ of actions, construct the progress formula
   $$Q \mapsto \bigvee_{u \in U} (Q \setminus {}^{\bullet}u) \cup u^{\bullet}.$$

46   Proof Graphs for Interleaved Progress

Leads-to properties can be proven by help of valid leads-to formulas that are picked up according to the pick-up rule of Sect. 45, and their combination according to Lemma 44.3.

Such proofs can conveniently be organized by means of *proof graphs*. The nodes of a proof graph are state formulas. The arcs starting at some node p represent a disjunction of leads-to formulas. A proof graph is acyclic, has one initial node, p, and one final node, q, thus proving $p \mapsto q$. As an almost trivial example, given $p \mapsto q$ and $q \mapsto r$, proof of the formula $p \mapsto r$ is represented by the proof graph

$$p \longmapsto q \longmapsto r . \tag{1}$$

Assuming standard notions of graphs, we define

46.1 Definition. *Let Σ be an es-net, let $p, q \in \mathrm{sf}(P_\Sigma)$ be state formulas and let G be a graph such that*

   i. G is directed, finite, and acyclic,
  ii. The nodes of G are state formulas in $\mathrm{sf}(P_\Sigma)$,
 iii. p is the only node without predecessor nodes,
 iv. q is the only node without successor nodes,
  v. for each node r, if $r_1, \ldots, r_n$ are the successor nodes of r, then $\Sigma \models r \mapsto (r_1 \vee \ldots \vee r_n)$.

Then G is a proof graph for $p \mapsto q$ in Σ.

The following theorem presents the central property of proof graphs:

46.2 Theorem. *Let Σ be an es-net and let G be a proof graph for $p \mapsto q$ in Σ. Then $\Sigma \models p \mapsto q$.*

Proof. The Theorem is shown by induction on the length n of a longest path in Σ. Induction basis: for $n = 1$, requirements iii and iv of Def. 46.1 imply $p = q$, hence $\Sigma \models p \mapsto q$ by Lemma 44.3(i).

For the induction step, let $a_1 \ldots a_n$ be the sequence of nodes of a longest path of G, and assume inductively the Theorem holds for each proof graph with longest paths of length $n - 1$.

For each successor node r of p in G, let G_r be the subgraph of G consisting of all nodes and arcs between r and q. G_r is a proof graph for $r \mapsto q$. Furthermore, the longest path of G_r has length $n - 1$, hence the inductive assumption implies $\Sigma \models r \mapsto q$.

Now, let $r_1, \ldots, r_m$ are the successor nodes of p. Then $\Sigma \models p \mapsto (r_1 \vee \ldots \vee r_m)$ by Def. 46.1(v). Furthermore, $\Sigma \models (r_1 \vee \ldots \vee r_m) \mapsto q$ by m-fold application of the above argument, and Lemma 44.3(iv). Hence $\Sigma \models p \mapsto q$ by Lemma 44.3(iii). $\qquad\square$

An arc from a node p to a node q is usually depicted $p \mapsto q$ (as in (1)). The special case of a progress set $U = \{u_1, \ldots, u_n\}$ and a property $Q \mapsto \bigvee_{u \in U} \mathrm{eff}(Q, u)$ picked up by Theorem 45.5 is frequently depicted as

$$
\begin{array}{c}
u_1 \quad q_1 \\
p \quad u_n \quad \vdots \\
q_n
\end{array}
\tag{2}
$$

A propositional implication $\Sigma \models p \rightarrow (q_1 \vee \ldots \vee q_n)$ is usually represented by

$$
\begin{array}{c}
q_1 \\
p \quad \vdots \\
q_n
\end{array}
\tag{3}
$$

A small proof graph for leads-to properties of $\Sigma_{38.1}$ exemplifies these conventions:

$$B \overset{b}{\mapsto} E \rightarrow EC \overset{e}{\mapsto} A \tag{4}$$

proves $\Sigma_{38.1} \models B \mapsto A$ as follows: $\Sigma_{38.1} \models B \mapsto E$ follows from Lemma 45.3 with the progress set $\{b\}$. The implication $\Sigma_{38.1} \models E \rightarrow EC$ follows from the place invariant $B + E - C = 0$, and $\Sigma \models EC \mapsto A$ follows again from Lemma 45.3 with the progress set $\{e\}$.

47   Standard Proof Graphs

The construction of a proof graph particularly includes determination of correct successor states of each node, as required in Def. 46.1(v). The pick-up rule 45.3 fortunately produces valid formulas that perfectly fit this purpose. Whenever the pick-up rule fails at some node r (as no action is enabled at r), place invariants and traps may specialize r by help of a valid implication $r \rightarrow r_1 \vee \ldots \vee r_n$. This again fits into the schema of proof graphs, according to Lemma 44.3(i). Proof graphs constructed in this way are called *standard proof graphs*:

47.1 Definition. *Let Σ be an es-net, let $p, q \in \mathrm{sf}(P_\Sigma)$ and let G be a proof graph for $p \mapsto q$ in Σ. G is a* standard proof graph *iff*

   i. Each node $r \neq q$ is a conjunction $r = r_1 \wedge \ldots \wedge r_n$ of atomic formulas $r_1, \ldots, r_n \in P_\Sigma$.

  ii. For each node r and its direct successor nodes $r_1, \ldots, r_n$ holds: Either $\Sigma \models r \rightarrow (r_1 \vee \ldots \vee r_n)$, or the pick-up rule yields $\Sigma \models r \mapsto (r_1 \vee \ldots \vee r_n)$.

In fact, all proof graphs of Sect. 46 are standard proof graphs. In addition to the properties and propositional implications, a question mark indicates an action that is not guarantees to be enabled, as for $\Sigma_{45.2}$:

$$A \begin{array}{c} \overset{a}{\nearrow} C \\ \underset{b?}{\searrow} D \end{array} \tag{1}$$

The motivating examples of Sect. 44 can now be proven by help of standard proof graphs. For example, Fig. 47.1 shows a standard proof graph for $\Sigma_{44.1} \models A \mapsto E$ and Fig. 47.2 shows a standard proof graph for $\Sigma_{44.2} \models AB \mapsto (F \vee DG)$.

As a slightly nontrivial example we construct a proof graph to show a central property of the asymmetrical *mutex* algorithm $\Sigma_{13.10}$: The *prepared* writer eventually gets *writing*. In terms of the representation of Fig. 47.3 we have to show

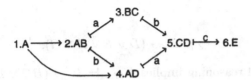

Figure 47.1. Standard proof graph for $\Sigma_{44.1} \models A \mapsto E$

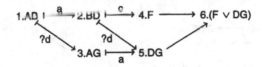

Figure 47.2. Standard proof graph for $\Sigma_{44.2} \models AB \mapsto (F \vee DG)$

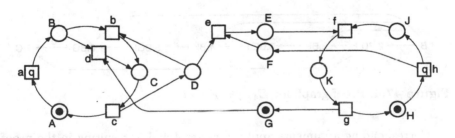

Figure 47.3. $\Sigma \models B \mapsto C$

$\Sigma_{47.3} \models B \mapsto C.$ \hfill (2)

The pick-up rule of Sect. 45 does not apply to B, because B itself is not progress prone. So we apply the invariant $D + E + K + G - A - B = 0$ which implies

$$\Sigma_{47.3} \models B \to (D \vee E \vee K \vee G).$$ \hfill (3)

Propositional reasoning implies $\Sigma_{47.3} \models B \to (BD \vee BE \vee BK \vee BG)$, and Lemma 44.3(i) furthermore yields $\Sigma_{47.3} \models B \mapsto (BD \vee BE \vee BK \vee BG)$. This is the justification for node 1 in the following first step of a proof graph:

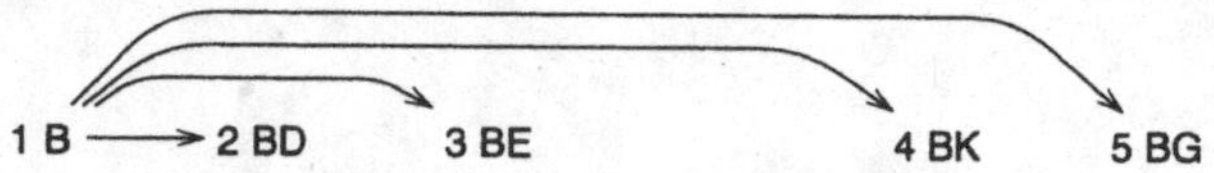

(4)

To continue the construction of a proof graph, we consider the nodes involved. Node 2, BD, enables action b. Hence Lemma 45.3 applies. The place invariant $C + D + E + K + G$, as outlined in Fig. 41.5, implies $D \to \neg G$, hence $\{B, D\}$ prevents d. This yields $BD \mapsto C \vee BE$. Graphically,

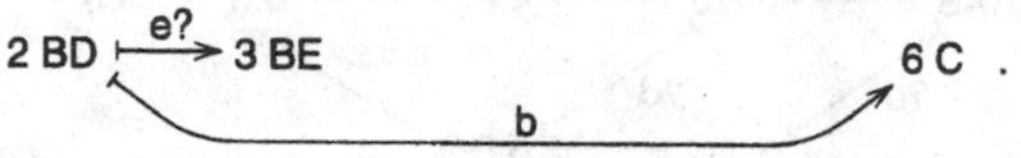

(5)

The question mark at "e" indicates that action e is not necessarily enabled. Node 3, BE, like node 1 enables none of the actions. Again, a place invariant helps: The place invariant $J - E - F = 0$ implies $\Sigma_{47.3} \models E \to J$, hence $\Sigma_{47.3} \models BE \to BEJ$, or graphically,

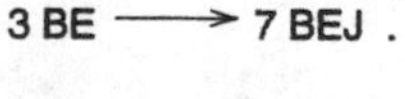

$$3\,BE \longrightarrow 7\,BEJ.$$ \hfill (6)

Node 7 now enables f. BEJ implies $\neg D$ and $\neg G$, hence prevents b and d, leaving change set $\{f\}$.

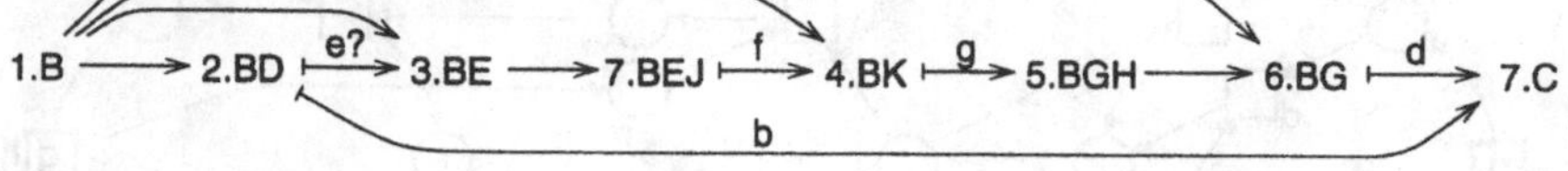

Figure 47.4. Proof graph for $\Sigma_{47.3} \models B \mapsto C$

Corresponding arguments apply to nodes 4 and 5, resulting in the proof graph shown in Fig. 47.4. $\Sigma_{47.3} \models DJ \mapsto (C \vee K)$ is shown likewise in Fig. 47.5. Its nodes are justified as follows: Node 1 by invariant $C + D - F + J + K + G = 1$. Node 2: DFJ prevents f, by invariant $F + E - J = 0$. Nodes 3–5: trivial.

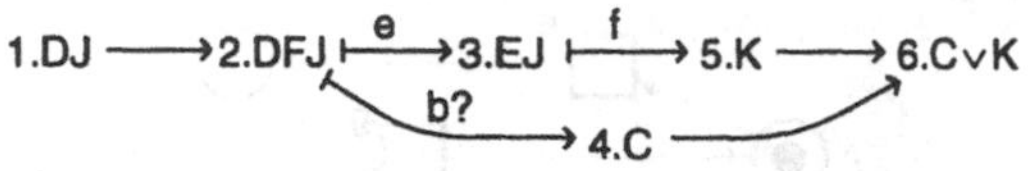

$$1.DJ \longrightarrow 2.DFJ \overset{e}{\longmapsto} 3.EJ \overset{f}{\longmapsto} 5.K \longrightarrow 6.C\lor K$$
$$\overset{b?}{\searrow} \longrightarrow 4.C \longrightarrow$$

Figure 47.5. Proof graph for $\Sigma_{47.3} \models DJ \mapsto C \lor K$

Standard proof graphs are easily understood and checked. However, there is no formalism to construct small and intuitive normal proof graphs. For example, loss of information is occasionally mandatory, as in the standard proof graph

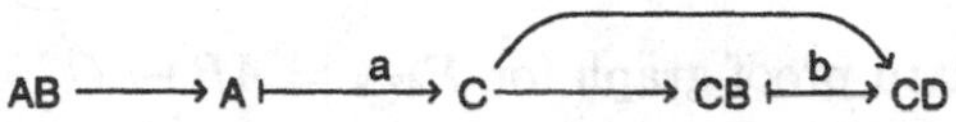

$$(8)$$

$$AB \longrightarrow A \overset{a}{\longmapsto} C \overset{\frown}{\longrightarrow} CB \overset{b}{\longmapsto} CD$$

proving $\Sigma_{47.6} \models AB \mapsto CD$.

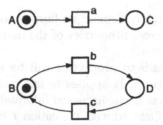

Figure 47.6. $\Sigma \models AB \mapsto CD$

Not each valid progress formula can be proven by a normal proof graph. An example is $\Sigma_{47.7} \models AB \mapsto C$. This deficit can be repaired by the complement E of B, as in Fig. 47.8. Then $\Sigma_{47.8} \models AB \mapsto C$ is proven by the normal proof graph of Fig. 47.9 (with place invariant $B + E = 1$ for node 3).

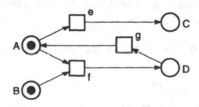

Figure 47.7. $\Sigma \models AB \mapsto C$ is not derivable by a standard proof graph

Whether or not each valid progress formula of an es-net can be proven by a standard proof graph together with complements, is left as an open problem.

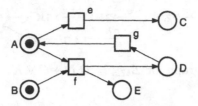

Figure 47.8. Extending $\Sigma_{47.7}$ by local state E, representing $\neg B$

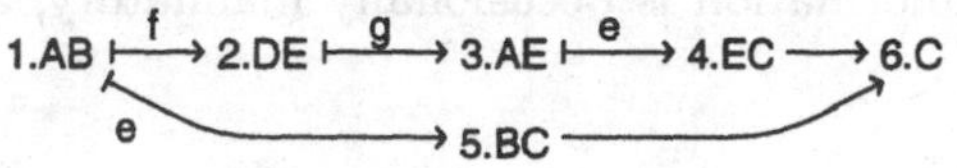

$$1.AB \overset{f}{\longmapsto} 2.DE \overset{g}{\longmapsto} 3.AE \overset{e}{\longmapsto} 4.EC \longrightarrow 6.C$$
$$\overset{e}{\longmapsto} 5.BC$$

Figure 47.9. Standard proof graph for $\Sigma_{47.8} \models AB \mapsto C$

48   How to Pick Up Fairness

Progress properties frequently depend on fairness assumptions. For example, none of the essential progress properties of the mutex algorithms in Sect. 13 holds if fairness is neglected.

A pick-up rule for leads-to properties will be given in this section, exploiting fairness assumptions. It applies to fair transitions that are *conflict reduced*. This property has been discussed informally in the introduction of Sect. 13 already: A conflict reduced transition t has at most *one* forward branching place in $^\bullet t$. In fact, almost all algorithms considered so far deal with fair transitions that are conflict reduced. The state testing mutex algorithm $\Sigma_{13.4}$ is the only exception.

48.1 Definition. *Let Σ be an es-net and let $t \in T_\Sigma$. t is* conflict reduced *iff there exists at most one $p \in {}^\bullet t$ with $p^\bullet \supsetneq \{t\}$. In this case, p is called the* conflict place *of t.*

48.2 Theorem. *Let Σ be an es-net and let $t \in T_\Sigma$ be fair and conflict reduced, with conflict place p. For $Q := {}^\bullet t \setminus \{p\}$ assume furthermore $\Sigma \models Q \mapsto p$. Then $\Sigma \models Q \mapsto t^\bullet$.*

Proof. Let $w = a_0 \overset{t_1}{\longrightarrow} a_1 \overset{t_2}{\longrightarrow} \dots$ be an interleaved run of Σ. For each Q-state a_k of w,

$$t_{k+1} = t \quad \text{or} \quad a_{k+1} \models Q \tag{1}$$

because t is conflict reduced. Furthermore, to each Q-state a_k there exists a p-state $a_{l'}$ with $l' \geq k$ (by the theorem's assumption of $\Sigma \models Q \mapsto p$). Let l be the smallest such index. Then for all $k < i \leq l$, $t_i \neq t$ (because $p \in {}^\bullet t$), hence $a_i \models Q$ (by (1), with induction on i), hence $a_l \models {}^\bullet t$.

Summing up, to each Q-state a_k there exists an index $l > k$ with

$$a_{l-1} \models {}^\bullet t \quad \text{and} \quad (t_l = t \quad \text{or} \quad a_l \models Q). \tag{2}$$

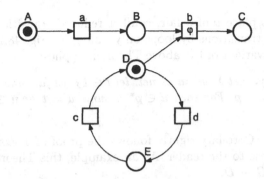

Figure 48.1. $\Sigma \models B \mapsto C$

To show $w \models Q \mapsto t^\bullet$, let a_k be any Q-state. Then there exists an index $l > k$ with $t_l = t$ or a sequence $l_0 < l_1 < \ldots$ of indices with $a_{l_i} \vdash {}^\bullet t$, for $i = 0, 1, \ldots$ (by (2) and induction on l). Then

(a) there exists an index $l > k$ with $t_l = t$ or
(b) w neglects fairness for t (by Def. 7.1(i))

Case (b) is ruled out by the theorem's assumption of fairness for t. Hence $w \models Q \mapsto t^\bullet$ by Def. 44.2(i). The theorem follows with Def. 44.2(ii).   □

As an example, action b of $\Sigma_{48.1}$ is fair and conflict reduced, with conflict place D. In order to show $({}^\bullet b \setminus \{D\}) \mapsto b^\bullet$, i.e.,

$$\Sigma_{48.1} \models B \mapsto C \tag{3}$$

we first show $\Sigma_{48.1} \models B \mapsto D$ by the proof graph

$$1.\text{B} \xrightarrow{} 2.\text{BE} \xmapsto{c} 3.\text{BD} \xrightarrow{} 4.\text{D} \tag{4}$$

where node 1 is based on the place invariant $A + B - D - E = 0$, and node 2 on the place invariant $C + D + E = 1$. Then the above theorem implies (3) (with $t = b$ and $Q = \{B\}$).

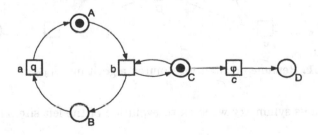

Figure 48.2. $\Sigma \models C \mapsto D$

The conflict place p of a fair, conflict reduced transition t is frequently linked to other transitions by loops only, as e.g. in Fig. 48.2 with $t = c$ and $p = C$. Then a variant of the above Theorem applies:

48.3 Corollary. *Let Σ be an es-net, let $t \in T_\Sigma$ be fair and conflict reduced, with conflict place p. For each $u \in p^\bullet$ assume $u = t$ or $u \in {}^\bullet p$. Then $\Sigma \models {}^\bullet t \mapsto t^\bullet$.*

Proof of this Corollary tightly follows the proof of Theorem 48.2, and is left as an exercise to the reader. As an example, this Theorem immediately yields $\Sigma_{48.2} \models C \mapsto D$.

49   Case Study:
Evolution of Mutual Exclusion Algorithms

We are now prepared to prove the evolution property of the mutual exclusion algorithms of Sect. 13. Evolution of the alternating algorithm $\Sigma_{13.3}$ is not guaranteed, and evolution of the state testing algorithm $\Sigma_{13.4}$ cannot be proven by means of Theorem 48.2, because the fair transitions b_l and b_r are not conflict reduced. The round-based algorithm $\Sigma_{13.6}$ is postponed to Sect. 56. Evolution of the asymmetrical algorithm $\Sigma_{13.10}$ has already been proven in Sect. 47. Evolution of all other algorithms of Sect. 13 will be proven in the sequel.

49.1   Evolution of the contentious algorithm

Figure 49.1 recalls the contentious algorithm, with renamed places. Due to

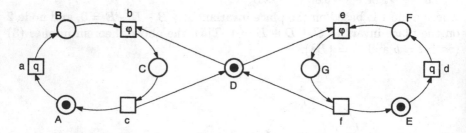

Figure 49.1. Renamed contentious mutex algorithm $\Sigma_{13.2}$

the algorithm's symmetry we stick to evolution of the left site, i.e.,

$$B \mapsto C. \tag{1}$$

The fair action b is conflict reduced, with conflict place D. First we show $B \mapsto D$ by means of the proof graph

$$\overset{\frown}{\text{1.B} \xrightarrow{\hspace{1cm}} \text{2.G} \overset{f}{\longmapsto} \text{3.D}} \tag{2}$$

Its node 1 is justified by the place invariant $A + B - D - G = 0$. Node 2 is trivial with Lemma 45.1.

Theorem 48.2 now immediately yields (1), with $t = b$, $p = D$ and $Q = \{B\}$.

49.2   Evolution of the token passing algorithm

The token passing algorithm of Fig. 13.5 is redrawn in Fig. 49.2 with renamed places. Due to the symmetry of the algorithm it suffices to show the evolution

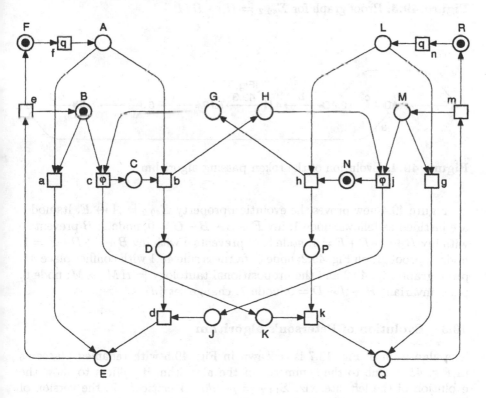

Figure 49.2. Renamed token passing mutex algorithm $\Sigma_{13.5}$

of the left site, viz $\Sigma_{13.5} \models pending_l \rightarrow critical_l$. In the version of Fig. 49.2 this reads $\Sigma_{49.2} \models A \mapsto E$.

In a separate calculation we first show $\Sigma_{49.2} \models H \mapsto HM$.

This property will be used twice: as part of the proof graph in Fig. 49.4, and as argument in the justification of one of its nodes, employing the fairness rule.

Figure 49.3 gives a proof graph for $\Sigma_{49.2} \models H \mapsto HM$. Its nodes are justified as follows: node 1: inv $C + H - K - Q - M = 0$; node 2: inv $G + K - P = 0$; node 3: P prevents j with $M + N + P + Q = 1$; node 4: Q prevents j with $M + N + P + Q = 1$.

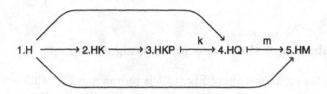

Figure 49.3. Proof graph for $\Sigma_{49.2} \models H \mapsto HM$

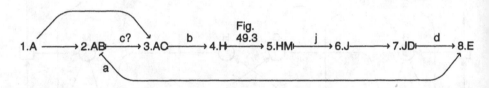

Figure 49.4. Evolution of the token passing algorithm

Figure 49.4 now proves the evolution property $\Sigma_{49.2} \models A \mapsto E$. Its nodes are justified as follows: node 1: inv $F + A - B - C = 0$; node 2: B prevents b with inv $B + C + D + E = 1$; node 3: C prevents a with inv $B + C + D + E = 1$; node 4: proof graph Fig. 49.2; node 5: fairness rule 48.1 with conflict place M, proof graph Fig. 49.2 and the propositional tautology $\models HM \rightarrow M$; node 6: place invariant $H + J - D = 0$; node 7: change set $\{d\}$.

49.3   Evolution of Peterson's algorithm

The algorithm of Fig. 13.7 is redrawn in Fig. 49.5 with renamed places as in Fig. 43.3. Due to the symmetry of the algorithm it suffices to show the evolution of the left site, viz. $\Sigma_{13.7} \models pend0_l \mapsto critical_l$. In the version of Fig. 49.5 this reads $\Sigma_{49.5} \models B \mapsto E$. Figure 49.6 gives a proof graph for this property.

The following place invariants will contribute to justify its nodes:

inv1: $A + B - F = 0$;
inv2: $G + H = 1$;
inv3: $L + M + N + P = 1$;
inv4: $C + D + E + F = 1$.

The nodes of Fig. 49.6 are justified as follows:

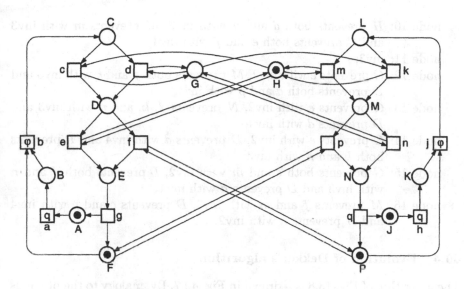

Figure 49.5. Renamed Peterson's mutex algorithm $\Sigma_{13.7}$

node 1: inv1;
node 2: Corollary 48.3;
node 3: inv2;
node 4: inv3;
node 5: H prevents d with inv2 and C prevents e with inv4;
node 6: G prevents c with inv2, M prevents k with inv3 and C prevents p with inv4;
node 7: G prevents c with inv2 and N prevents both k and n with inv3;
node 8: G prevents c with inv2 and P prevents both k and n with inv3;
node 9: G prevents both c and m with inv2 and L prevents n with inv3;

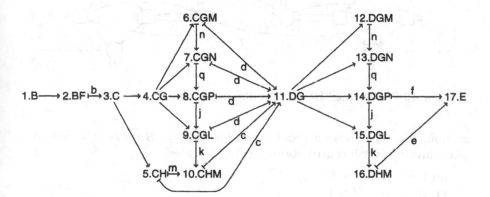

Figure 49.6. Evolution of Peterson's algorithm

node 10: H prevents both d and n with inv2, M prevents m with inv3 and C prevents both e and p with inv4;

node 11: inv3;

node 12: G prevents e with inv2, M prevents both f and k with inv3 and D prevents both d and p with inv4;

node 13: G prevents e with inv2, N prevents f, k, and n with inv3 and D prevents d with inv4;

node 14: G prevents e with inv2, D prevents d with inv4 and P prevents both k and n with inv3;

node 15: G prevents both e and m with inv2, L prevents both f and n with inv3 and D prevents d with inv4;

node 16: M prevents f and m with inv3, D prevents c and p with inv4 and H prevents n with inv2.

49.4   Evolution of Dekker's algorithm

The algorithm of Fig. 13.8 is redrawn in Fig. 49.7. By analogy to the previous

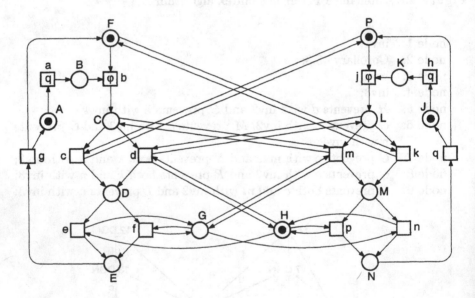

Figure 49.7. Renamed Dekker's mutex algorithm $\Sigma_{13.8}$

section, Fig. 49.8 shows a proof graph for $\Sigma_{49.7} \models B \mapsto D$. The following place invariants will contribute to justify its nodes:

inv1: $A + B - F = 0$;

inv2: $G + H = 1$;

inv3: $L + M + N + P = 1$;

inv4: $C + D + E + F = 1$.

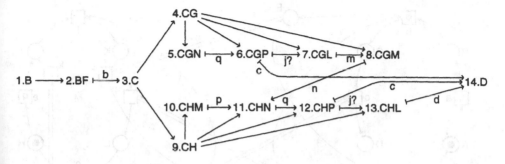

Figure 49.8. Evolution of Dekker's algorithm

The nodes of Fig. 49.8 are justified as follows:

node 1:   inv1;
node 2:   Corollary 48.3;
node 3:   inv2;
node 4:   inv3;
node 5: N prevents c, d, m, and n with inv3, C prevents f with inv4:
node 6: P prevents d, m, and n with inv3 and C prevents f with inv4;
node 7: L prevents c and n with inv3, G prevents d with inv2 and C prevents both f and k with inv4;
node 8: M prevents c, d, and m with inv3, G prevents p with inv2 and C prevents f with inv4;
node 9:   inv3;
node 10: M prevents c, m, and d with inv3, H prevents n with inv2 and C prevents e with inv4;
node 11: N prevents c, d, m, and p with inv3, C prevents e with inv4;
node 12: P prevents d, m, and p with inv3 and C prevents e with inv4;
node 13: L prevents both c and p with inv3, H prevents m with inv2 and C prevents both e and k with inv4.

49.5   Evolution of Owicki/Lamport's asymmetrical mutex

The algorithm of Fig. 13.9 is redrawn in Fig. 49.9. We will show different properties of the writer and the reader site, respectively. First we show that the pending writer will eventually be writing; formally: $\Sigma_{13.9} \models prep1 \mapsto writing$. In the version of Fig. 49.9 this reads $\Sigma_{49.9} \models B \mapsto D$. Figure 49.10 gives a proof graph for this property.

   The following invariants will contribute to justify its nodes:

inv1: $A + B - F = 0$;
inv2: $H + J + K + L + M = 1$;
inv3: $C + D - E = 0$;
inv4: $G + K + L = 1$;

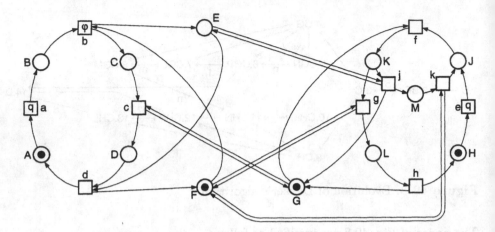

Figure 49.9. Renamed Owicki/Lamport's asymmetrical mutex algorithm $\Sigma_{13.9}$

inv5: $A + B + C + D = 1$;
inv6: $C + D + F = 1$;
inv7: $E + F = 1$;

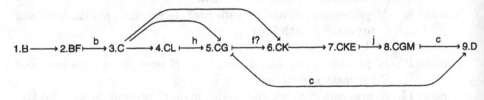

Figure 49.10. Evolution of the writer in Owicki/Lamport's algorithm

The nodes of Fig. 49.10 are justified as follows:

node 1: inv1;
node 2: Corollary 48.3;
node 3: inv4;
node 4: L prevents c by inv4;
node 5: trivial;
node 6: inv3;
node 7: K prevents c by inv4; E prevents g by inv7; C prevents d by inv5;
node 8: M prevents f by inv2 and C prevents k by inv6.

If the reader is pending, the algorithm guarantees that eventually the reader will be reading or the writer will be writing; formally: $\Sigma_{13.9} \models$ $pend1 \mapsto (writing \vee reading)$. In the version of Fig. 49.9 this reads $\Sigma_{49.9} \models$ $J \mapsto L \vee D$.

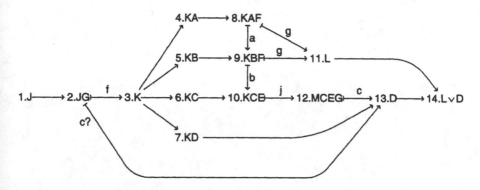

Figure 49.11. Evolution of the reader in Owicki/Lamport's algorithm

Figure 49.11 shows a proof graph for this property. In addition to the above invariants inv1, ..., inv7, the following invariant will contribute to justify the nodes of this proof graph:

inv8: $H + J - G = 0.$

The nodes of Fig. 49.11 are justified as follows:

node 1:   inv8;
node 2:   trivial;
node 3:   inv5;
node 4:   inv1;
node 5:   inv1;
node 6:   inv3;
node 7:   trivial;
node 8: K prevents k by inv2; A prevents b by inv5 and F prevents j by inv7;
node 9: K prevents k by inv2 and F prevents j by inv7;
node 10: K prevents c by inv4, C prevents d by inv5 and E prevents g by inv7;
node 11: trivial;
node 12: M prevents both j and f by inv2; C prevents d by inv5 and E prevents k by inv7;
node 13: trivial.

IX. Concurrent Progress of Elementary System Nets

The interleaving based *leads-to* operator $\mapsto$, considered in Chap. VIII, adequately describes important properties of a wide range of distributed algorithms. But a variety of progress properties, typical and specific for distributed algorithms, are not expressible by this operator. This particularly includes *rounds*, as informally described in the case studies of Chap. III.

A new operator "$\hookrightarrow$" will be introduced in Sect. 50 with $p \hookrightarrow q$ ("p causes q") interpreted over *concurrent* runs K: In K holds $p \hookrightarrow q$ iff each reachable p-state of K is followed by a reachable q-state.

50 Progress on Concurrent Runs

As an introductory example we return to the producer/consumer system $\Sigma_{8.1}$. Its behavior was intuitively described as a sequence of *rounds*, with each round consisting of an item's production, delivery, removal, and consumption. Each such round starts and ends in the initial state. The rounds of a run of $\Sigma_{8.1}$ are depicted in Fig. 50.1. We shall present means to represent and to reason about rounds of this kind and other progress properties based on concurrent runs. We start with syntax corresponding to Def. 44.1:

50.1 Definition. *Let P be a set of symbols and let $p, q \in \mathrm{sf}(P)$ be state formulas over P. Then the symbol sequence $p \hookrightarrow q$ ("p causes q") is a causes formula over P. The set of all such formulas will be denoted $\mathrm{cf}(P)$.*

Causes formulas are interpreted over concurrent runs and over es-nets:

50.2 Definition. *Let Σ be an es-net and let $p \hookrightarrow q \in \mathrm{cf}(P)$ be a causes formula.*

  i. *$p \hookrightarrow q$ is said to hold in a concurrent run K of Σ (written $K \models p \hookrightarrow q$) iff to each reachable p-state C of K there exists a q-state D of K that is reachable from C.*
  ii. *$p \hookrightarrow q$ is said to hold in Σ (written $\Sigma \models p \hookrightarrow q$) iff $K \models p \hookrightarrow q$ for each concurrent run K of Σ.*

As an example, the formulas $BC \hookrightarrow BE$ and $A \hookrightarrow CD$ both hold in $\Sigma_{50.2}$ (whereas $BC \mapsto BE$ and $A \mapsto CD$ don't hold).

The following lemma resembles Lemma 44.3:

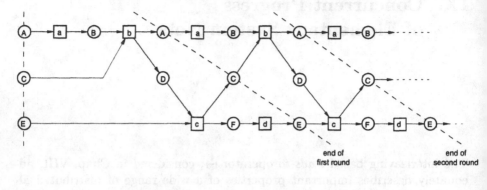

A : ready to produce	a : produce
B : ready to deliver	b : deliver
C : buffer empty	c : remove
D : buffer filled	d : consume
E : ready to remove	
F : ready to consume	

Figure 50.1. Rounds in the infinite run of $\Sigma_{8.1}$. Inscriptions as in Fig. 5.5

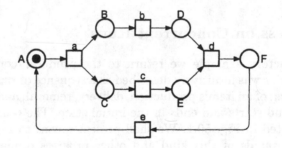

Figure 50.2. $\Sigma \models BC \hookrightarrow BE$ and $\Sigma \models A \hookrightarrow CD$

50.3 Lemma. *Let Σ be an es-net and let $p, q, r \in \mathrm{sf}(P_\Sigma)$.*

i. $\Sigma \models p \hookrightarrow p$;
ii. *If $\Sigma \models p \hookrightarrow q$ and $\Sigma \models q \hookrightarrow r$ then $\Sigma \models p \hookrightarrow r$;*
iii. *If $\Sigma \models p \hookrightarrow r$ and $\Sigma \models q \hookrightarrow r$ then $\Sigma \models (p \vee q) \hookrightarrow r$.*

In general, *causes* is weaker then *leads-to*. They coincide in special cases:

50.4 Lemma. *Let Σ be an es-net and let $p, q, r \in \mathrm{sf}(P_\Sigma)$.*

i. *If $\Sigma \models p \mapsto q$ then $\Sigma \models p \hookrightarrow q$;*
ii. *Let $Q \subseteq P_\Sigma$ and let $q = \bigvee Q$. If $\Sigma \models p \hookrightarrow q$ then $\Sigma \models p \mapsto q$.*

51   The Concurrent Pick-up Rule

Section 50 introduced means to represent causes properties. Means to *prove* such properties will be derived in the sequel. Valid causes formulas can be picked up from the static structure of a net. A corresponding *pick-up rule* will be based on *change sets*, as introduced for the leads-to-operator in Sect. 45.6.

The forthcoming pick-up rule highlights one distinguished feature: Picked up causes formulas $p \hookrightarrow q$ can be embedded into a context, r, yielding

$$r \wedge p \hookrightarrow r \wedge q. \tag{1}$$

First we consider a special case of the forthcoming most general pick-up rule, in Theorem 51.1.

As an example, from an es-net Σ with a part

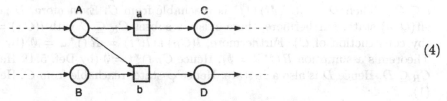
$$\tag{2}$$

the property $\Sigma \models A \hookrightarrow B$ can be picked up immediately. This in turn can be embedded into the context of any local state C, yielding

$$\Sigma \models CA \hookrightarrow CB. \tag{3}$$

As a more general example, from

$$\tag{4}$$

$\Sigma \models AB \hookrightarrow BC \vee D$ can be picked up immediately. This again can be embedded into the context of any local state E, yielding

$$EAB \hookrightarrow EBC \vee ED. \tag{5}$$

Abstractly formulated, let $Q \subseteq P_\Sigma$ be progress prone and let $U \subseteq Q^\bullet$ be a change set of Q. Then $\Sigma \models Q \hookrightarrow \bigvee_{u \in U} \text{eff}(Q, u)$. This of course resembles the pick-up rule for the leads-to operator $\hookrightarrow$, as stated in Lemma 45.3. But in contrast to picked up *yields* formulas, the above *causes* formula can be embedded in a context $R \subseteq P_\Sigma$, yielding

$$\Sigma \models R \cup Q \hookrightarrow R \cup (\bigvee_{u \in U} \text{eff}(Q, u)), \tag{6}$$

provided $^\bullet U \subseteq Q$ and $Q \cap R = \emptyset$.

Rule (6) suffices in most cases, and will be considered in Corollary 51.2. Rule (6) is occasionally too strict, as the following example shows: In (4), $\{A\}$ is progress prone and $\{a, b\}$ is a change set of $\{A\}$. Hence $\Sigma \models A \hookrightarrow C \vee D$ can be picked up immediately. But a context cannot be applied to this formula by means of (6) because $^\bullet\{a, b\} \not\subseteq \{A\}$. So, in (6) we skip the requirement of $^\bullet U \subseteq Q$, but allow context to $\text{eff}(Q, u)$ only in case $^\bullet u \subseteq Q$. For example, in (4) the formula $A \hookrightarrow C \vee D$ can now be embedded into the context of any local state E for the occurrence of a (because $^\bullet a \subseteq \{A\}$), but not for the occurrence of b (because $^\bullet b \not\subseteq \{A\}$), yielding

$$\Sigma \models EA \hookrightarrow EC \vee D. \tag{7}$$

Generally formulated, the change set U of a progress prone set Q is partitioned into $U = V \vee W$ such that $^\bullet V \subseteq Q$. Then a context R is applied to V only. Hence the following theorem:

51.1 Theorem. *Let Σ be an es-net, let $Q \subseteq P_\Sigma$ be progress prone and let $U = V \cup W$ be a change set of Q with $^\bullet V \subseteq Q$. Furthermore, let $R \subseteq P_\Sigma$ with $R \cap {}^\bullet V = \emptyset$. Then $\Sigma \models R \cup Q \hookrightarrow (R \cup \bigvee_{u \in V} \mathrm{eff}(Q, u)) \vee (\bigvee_{u \in W} \mathrm{eff}(Q, u))$.*

Proof. Let K be a concurrent run of Σ and let C be a reachable $R \cup Q$-state. With $\varphi := (R \cup \bigvee_{u \in V} \mathrm{eff}(Q, u)) \vee (\bigvee_{u \in W} \mathrm{eff}(Q, u))$ we have to show:

$$K \text{ has a } \varphi\text{-state that is reachable from } C. \tag{1}$$

There exists a subset $C_Q \subseteq C$ with $l(C_Q) = Q$. Then $l(C_Q)$ enables at least one progressing action u (by the theorem's assumption that Q is progress prone). Then $C_Q \not\subseteq K^\circ$ (by Def. 8.2(ii)). Furthermore, $u \in U$ (as U is a change set of Q). Then there exists some $t \in C_Q{}^\bullet$ with $l(t) = u$ (by Def. 5.4(ii)).

If $u \in V$, then $^\bullet u \subseteq Q$ (by the theorem's assumption $^\bullet V \subseteq Q$), hence $^\bullet t \subseteq C_Q$. Then $D := (C \setminus {}^\bullet t) \cup t^\bullet$ is reachable from C. Even more, D is an $\mathrm{eff}(Q, u)$-state. Furthermore, there exists a subset $C_R \subseteq C$ with $l(C_R) = R$ (by construction of C). Furthermore, $l(C_R) \cap l(^\bullet t) = R \cap {}^\bullet u = \emptyset$ (by the Theorem's assumption $R \cap {}^\bullet U = \emptyset$). Hence $C_R \cap {}^\bullet t = \emptyset$ (by Def. 5.4). Hence $C_R \subseteq D$. Hence D is also a R-state. Hence φ-state, reachable from C. Hence (1).

In case of $u \in W$, let t' be a minimal (with respect to $<_K$) element with $t' \in C_Q{}^\bullet$ and $l(t) = u$. Then there exists a state E, reachable from C, with $C \cup {}^\bullet t \subseteq E$. Then $F := (E \setminus {}^\bullet t') \cup t'^\bullet$ is an $\mathrm{eff}(Q, u)$-state, reachable from C, hence (1). □

According to this theorem, in fact (7) is valid in (4). The following special case with $W = \emptyset$ (hence $V = U$) suffices in most cases (e.g., for the validity of (5) and (4)).

51.2 Corollary. *Let Σ be an es-net, let $Q \subseteq P_\Sigma$ be progress prone and let $U \subseteq T_\Sigma$ be a change set of Q with $^\bullet U \subseteq Q$. Furthermore, let $R \subseteq P_\Sigma$ with $R \cap Q = \emptyset$. Then $\Sigma \models R \cup Q \hookrightarrow R \cup \bigvee_{u \in U} \mathrm{eff}(Q, u)$.*

The opposite special case (i.e., $V = R = \emptyset$) mirrors the interleaved pick-up rule.

52   Proof Graphs for Concurrent Progress

Picked-up causes formulas can be composed in *proof graphs*. The successor nodes $r_1, \ldots, r_n$ of a node r then represent $r \hookrightarrow (r_1 \vee \ldots \vee r_n)$. All other aspects of such proof graphs coincide with proof graphs for leads-to formulas:

52.1 Definition. *Let Σ be an es-net, let $p, q \in \text{sf}(P_\Sigma)$ be state formulas, and let G be a graph meeting Def. 46.1(i)–(iv) and*

vi. for each node r, if $r_1, \ldots, r_n$ are the successor nodes of r, then $\Sigma \models r \hookrightarrow (r_1 \vee \ldots \vee r_n)$.

Then G is a proof graph for $p \hookrightarrow q$ in Σ.

Proof graphs for causes formulas in fact prove validity of those formulas:

52.2 Theorem. *Let Σ be an es-net and let G be a proof graph for $p \hookrightarrow q$ in Σ. Then $\Sigma \models p \hookrightarrow q$.*

Proof of this theorem is essentially the same as the proof of Theorem 46.2 and is left as an exercise to the reader.

Standard proof graphs are constructed from propositional implications and picked-up formulas:

52.3 Definition. *Let Σ be an es-net, let $p, q \in \text{sf}(P_\Sigma)$, and let G be a proof graph for $p \hookrightarrow q$ in Σ. G is a standard proof graph iff*

i. Each node $r \neq q$ is a conjunction $r = r_1 \wedge \ldots \wedge r_n$ of atomic formulas $r_1, \ldots, r_n \in P_\Sigma$.

ii. For each node r and its direct successor nodes $r_1, \ldots, r_n$, either $\Sigma \models r \rightarrow (r_1 \vee \ldots \vee r_n)$ or the pick-up rule Theorem 51.1 yields $\Sigma \models r \hookrightarrow (r_1 \vee \ldots \vee r_n)$.

Leads-to properties $p \mapsto q$ can frequently be proven by means of short proof graphs for $p \hookrightarrow q$, together with Lemma 50.4(ii). For example, the property $\Sigma_{44.1} \models A \mapsto E$, as proven in Fig. 47.1, can likewise be proven by means of the – shorter – proof graph in Fig. 52.1 for $\Sigma_{44.1} \models A \hookrightarrow E$, and Lemma 50.4(ii). The forthcoming concept of *rounds* provides further means for short proof of both *causes* and *leads-to* formulas.

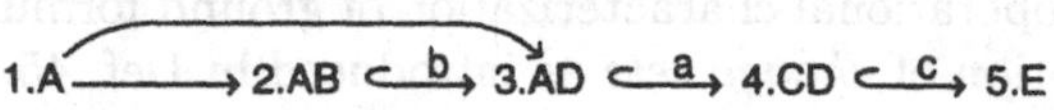

$$1.A \longrightarrow 2.AB \xrightarrow{\ b\ } 3.AD \xrightarrow{\ a\ } 4.CD \xrightarrow{\ c\ } 5.E$$

Figure 52.1. Standard proof graph for $\Sigma_{44.1} \models A \hookrightarrow E$

53 Ground Formulas and Rounds

A state formula $p \in \text{sf}(P_\Sigma)$ of an es-net Σ is said to be a *ground formula of Σ* if in each concurrent run, each reachable state of Σ is followed by a p-state; formally

$$\Sigma \models \text{true} \hookrightarrow p. \tag{1}$$

Distributed algorithms can frequently be properly comprehended and verified using ground formulas. Interesting ground formulas are mostly conjunctions (viz. subsets) of atoms $p \subseteq P_\Sigma$, or even distinguished reachable states. Such a state is said to be a *ground state*. For example, the initial state ACE of the producer/consumer system $\Sigma_{8.1}$, as redrawn in $\Sigma_{53.1}$, is a ground state, and in fact the only ground state of $\Sigma_{53.1}$.

Claim (1) implies that each finite run of Σ ends at a p-state, and that each infinite run of Σ has infinitely many p-states. Distributed algorithms are frequently *round-based*. A *round* of an es-net Σ is a finite, Σ-based concurrent run that starts and ends at a ground state. Σ is *round-based* if there exists a finite set R of rounds such that each concurrent run K of Σ can be conceived as a finite or infinite sequence of rounds of R. As an example, there exists a unique round of $\Sigma_{53.1}$, as outlined in Fig. 50.1.

53.1 Definition. *Let Σ be an es-net and let $p \in \mathrm{sf}(P_\Sigma)$ be a state formula. p is a* ground formula *of Σ iff $\Sigma \models \mathrm{true} \hookrightarrow p$.*

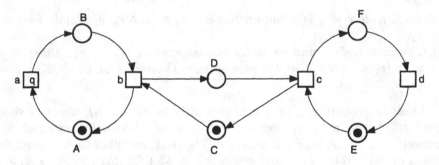

Figure 53.1. Ground formula ACE

There is an operational characterization of ground formulas $p \subseteq P_\Sigma$. It is based on the notion of *change sets* as introduced in Def. 45.4(ii) already.

53.2 Theorem. *Let Σ be an es-net and let $p \subseteq P_\Sigma$ with*

   i. *$\Sigma \models a_\Sigma \hookrightarrow p$, and*
   ii. *for some change set U of p and each $u \in U$, $\Sigma \models \mathrm{eff}(p, u) \hookrightarrow p$.*

Then p is a ground formula.

Proof. Let K be a concurrent run of Σ, and let $A, B \subseteq P_K$ be two reachable states of K.

A transition $t \in T_K$ is said to be *between* A and B iff for some $a \in A$ and some $b \in B$ holds: $a \leq_K t \leq_K b$. Let $dist(A, B)$ denote the set of all transitions between A and B. Obviously $dist(A, B) = \emptyset$ iff B is reachable from A.   (2)

The proof essentially bases on the following proposition:

. Let $dist(A, B) \neq \emptyset$ and let A be a p-state. Then there exists a (3)
reachable p-state D of K with $dist(D, B) \subsetneq dist(A, B)$.

This proposition is proven as follows: $dist(A, B) \neq \emptyset$ implies a transition
$t \in A^\bullet$ with $t < b$ for some $b \in B$. Then there exists a transition u with
$^\bullet u \subseteq A$ and $u < b$, because t has only finitely many predecessors in K.
Let $C := (A \setminus {}^\bullet u) \cup u^\bullet$. Then $dist(C, B) = dist(A, B) \setminus \{u\}$. If C is a p-
state, we are done (with $D := C$). Otherwise $l(u) \in p^\bullet$, hence $l(u) \in U$ for
each change set U of p. Furthermore $l(C)$ is an eff(A, u)-state of Σ. Then
$\Sigma \models l(C) \hookrightarrow p$ by the Theorem's assumption. Then there exists a p-state D
of K, reachable from C. Furthermore, $dist(A, B) \subsetneq dist(C, B) \subseteq dist(D, B)$.
Hence the proposition (2).

Now let B be any reachable state of K. The Theorem's assumption $\Sigma \models$
$a_\Sigma \hookrightarrow p$ implies a reachable p-state, A of K. A $dist(A, B)$ is finite, finitely
many applications of the proposition (2) yield a reachable p-state D of K
with $dist(D, B) = \emptyset$. Then (1) implies D be reachable from B hence the
Theorem.   □

As an example, we prove that the initial state ACE is a ground formula
of $\Sigma_{53.1}$ by means of Theorem 53.2. The first condition, $\Sigma \models a_\Sigma \hookrightarrow ACE$,
is trivially fulfilled with Lemma 50.3(i). For the second condition of Theo-
rem 53.2 observe that $U = \{a\}$ is a change set of ACE, because A prevents
b by inv $A + B = 1$ and C prevents c by inv $C + D = 1$. Hence we have to
show: $\Sigma_{53.1} \models BCE \hookrightarrow ACE$. The proof graph

$$1.BCE \overset{b}{\hookrightarrow} 2.ADE \overset{c}{\hookrightarrow} 3.ACF \overset{d}{\hookrightarrow} 4.ACE \qquad (4)$$

shows this property. Its nodes are justified as follows:

    node 1: context E;
    node 2: context A;
    node 3: context AC.

Hence (4) proves that ACE will eventually be reached from *any* reachable
state, though (4) does not refer to all reachable states of $\Sigma_{53.1}$, and ignores,
e.g., BDE or BDF!

As a further, technical example we show that the initial state AD of $\Sigma_{53.2}$
is a ground state: According to Theorem 53.2 it suffices to show that

$$BD \hookrightarrow AD, \quad \text{and} \qquad (5)$$

$$AE \hookrightarrow AD \qquad (6)$$

both hold in $\Sigma_{53.2}$, as $\{a, d\}$ is a change set of AD. Figure 53.3 shows a proof
graph for (5). Its nodes are justified as follows:

    node 1: Theorem 51.1, with $V = \{b\}$, $W = \{g\}$, $R = \{D\}$;
    node 2: context F;

node 3: context A;
node 4: context D.

Corollary 51.2 was not sufficient to justify node 1.

Proof of (6) is left to the reader, due to the symmetrical structure of $\Sigma_{53.2}$.

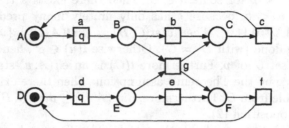

Figure 53.2. AD is a ground state

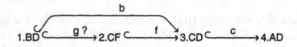

Figure 53.3. $\Sigma_{53.2} \models BD \hookrightarrow AD$

Ground formulas support the proof of any causes formulas:
In Theorem 51.1, the requirement that Q is progress prone may be replaced
by the requirement to imply $\neg p$ for some ground formula p. An element t of
the change set U with $^\bullet t \subseteq Q$ is obsolete in this case. Hence:

53.3 Theorem. *Let Σ be an es-net and let p be a ground formula of Σ. Let
$Q \subseteq P_\Sigma$ with $\Sigma \models Q \to \neg p$ and let U be a change set of Q in Σ. Then
$\Sigma \models Q \hookrightarrow \bigvee_{u \in U} \text{eff}(Q, u)$.*

Proof. Let K be a concurrent run of Σ, let C be a Q-state of K, and let
$C_Q \subseteq C$ with $l(C_Q) = Q$. Then there exists a p-state D of K that is reachable
from C, because p is a ground formula. From $\Sigma \models Q \to \neg p$ follows $C_Q \not\subseteq D$.
Hence there exists a transition $t \in C_Q^\bullet$ in K, with $l(t) \in U$. Hence the
proposition.   □

As an example we show that the filled buffer of the producer/consumer
system $\Sigma_{53.1}$ will eventually be empty:

$$\Sigma_{53.1} \models D \hookrightarrow C. \tag{7}$$

Based on the above proven ground formula ACE we apply Theorem 53.3 as
follows: The buffer is filled in some state $a \subseteq P_{\Sigma_{53.1}}$ iff $D \in a$. Furthermore,
$\Sigma_{53.1} \models D \to \neg C$ by inv $D + C = 1$; hence $\Sigma_{53.1} \models D \to \neg ACE$. $U = \{c\}$

is a change set of D and eff$(D, c) = CF$. Hence with Theorem 53.3: $\Sigma_{53.1} \models D \hookrightarrow C$, hence (7) with Lemma 50.4(ii).

54 Rounds of Sequential and Parallel Buffer Algorithms

54.1 Rounds of the sequential two-cell algorithm

The initial state ACE of the basic producer/consumer algorithm $\Sigma_{8.1}$, as redrawn in Fig. 53.1, is a ground state, i.e., a ground formula that even is a reachable state of $\Sigma_{53.1}$. This has been proven in Sect. 53 already, and has been outlined in Fig. 50.1. The sequential buffer with two cells, as outlined in Fig. 9.1, has a unique ground state, too, i.e., its initial state $ACEG$. Proof of this property strictly follows the corresponding proof of $\Sigma_{53.1}$, and is left as an exercise to the reader.

54.2 A ground formula of the nondeterministic parallel algorithm

Figure 9.2 has no ground state. As Fig. 9.5 exemplifies, one of the buffer cells may remain filled forever. However, the algorithm has a ground formula, AG, indicating that the producer always returns to *ready to produce*, and the consumer to *ready to remove*. This property can easily be proven by means of Theorem 53.2: $a_{\Sigma_{9.2}} \rightarrow AG$ is a propositional tautology, hence trivially $a_{\Sigma_{9.2}} \hookrightarrow AG$. Furthermore, $\{a, d, e\}$ is a change set of AG, hence we have to show

$$BG \hookrightarrow AG, \quad AEH \hookrightarrow AG, \quad \text{and} \quad ACH \hookrightarrow AG. \tag{1}$$

The first of those propositions follows from the standard proof graph

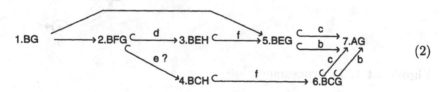

$$\tag{2}$$

Justification of its nodes as well as proof of the rest of (1) is left as an exercise to the reader.

54.3 A ground formula of the deterministic parallel algorithm

The deterministic parallel producer/consumer of Fig. 9.3 has likewise no ground state: Some of its finite runs terminate in the initial state, and some terminate in the state $ACEGKM$. But $ACEG$ is a ground formula of $\Sigma_{9.3}$, indicating the producer and the consumer *ready* and both buffer cells *empty*.

This again can be shown using the Theorem 53.2: $a_{\Sigma_{9.3}} \to ACEG$ is a propositional tautology. $\{a\}$ is a change set of $ACEG$, because A prevents b and c by inv $A + B = 1$, E prevents d by inv $E + F = 1$, and C prevents e by inv $C + D = 1$. Hence one has to show $\Sigma_{9.3} \models BCEG \hookrightarrow ACEG$. This can be achieved by means of a standard proof graph, left as an exercise to the reader.

54.4   Rounds of the two consumers algorithm

Figure 54.1 shows a system with *two* consumers. Its initial state is a ground state. This follows from Theorem 53.2 by means of the proof graph

$$1.\ BCEG \xrightarrow{b} 2.\ ADEG \quad\quad\quad 5.\ ACEG \tag{3}$$

with intermediate states 3. $ACFG$ (via c), 4. $ACHE$ (via d, e), and f.

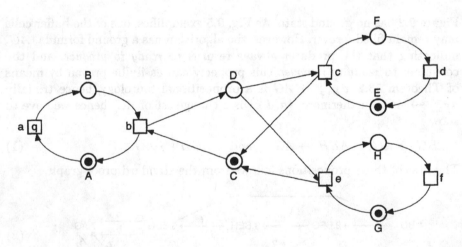

Figure 54.1. Two consumers

As a variant one may turn $\Sigma_{54.1}$ into a deterministic algorithm, serving the consumers alternately. This is easily achieved by means of the synchronization circuit

$$\tag{4}$$

augmented to $\Sigma_{54.1}$. The above ground formula $ACEG$ then is no longer a ground state, but only a ground formula.

55 Rounds and Ground Formulas of Various Algorithms

Virtually all algorithms of Chap. II are round-based or at least have ground formulas. Being aware of rounds and ground formulas, the reader obtains better intuitive perception of an algorithm. In this section we discuss rounds of various algorithms introduced in Chap. II, including the philosophers, the asynchronous pushdown, and the crosstalk algorithm. Rounds and ground formulas of mutual exclusion algorithms are postponed to Sect. 56.

55.1 Rounds of the philosophers algorithm

The algorithm for thinking and eating philosophers, as given in Fig. 10.1, operates in five rounds, one for each philosopher. Each round comprises an instance of picking up and releasing a philosopher's forks, as outlined in (2) of Sect. 10.

Upon proving this property, we first observe that the initial state σ of $\Sigma_{10.1}$ enables five actions, A_p, B_p, C_p, D_p, and E_p, i.e., each philosopher is able to pick up his forks. In fact, $U = \{A_p, \ldots, E_p\}$ is a change set of σ. Let $\sigma_A, \ldots, \sigma_E$ be the states reached after the occurrence of $A_p, \ldots, E_p$, respectively. Hence there are five steps $\sigma \xrightarrow{A_p} \sigma_A, \ldots, \sigma \xrightarrow{E_p} \sigma_E$ starting at σ. In order to show that σ is in fact a ground state, with Theorem 53.2 we have to show $\sigma_A \hookrightarrow \sigma, \ldots, \sigma_E \hookrightarrow \sigma$. This in turn is almost trivial, because one can pick up $\sigma_A \xrightarrow{A_r} \sigma, \ldots, \sigma_E \xrightarrow{E_r} \sigma$ immediately, according to Corollary 51.2 (with empty context).

As a consequence, each concurrent run has a linearization (in general not unique) that consists of a sequence of eating cycles, as represented in (2) and (3) of Sect. 10.

55.2 Rounds of the asynchronous pushdown algorithm

The algorithm that organizes control in an asynchronous stack with capacity for four items has been given in Fig. 11.2, and is redrawn in Fig. 55.1, with renamed places. It operates in two rounds, one to push a value into the stack, and one to pop a value. Each round comprises either an entire "wave" of pushing down data along the actions $a_0, \ldots, a_4$, or an entire wave of soliciting data along $b_0, \ldots, b_4$.

Upon proving this fact, we first observe that the initial state σ of $\Sigma_{55.1}$ enables two actions, a_0 and b_0, i.e., the initial state is enabled for both a *push* round and a *pop* round. In fact, $U = \{a_0, b_0\}$ is a change set of σ. Hence two

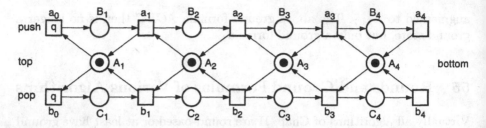

Figure 55.1. Renamed asynchronous stack $\Sigma_{11.2}$

steps $\sigma \xrightarrow{a_0} B_1 A_2 A_3 A_4$ and $\sigma \xrightarrow{b_0} C_1 A_2 A_3 A_4$ start at σ. In order to show that σ is a ground state, with Theorem 53.2 we have to show $B_1 A_2 A_3 A_4 \hookrightarrow \sigma$ and $C_1 A_2 A_3 A_4 \hookrightarrow \sigma$. This is easily achieved by two proof graphs

$$B_1 A_2 A_3 A_4 \xrightarrow{a_1} A_1 B_2 A_3 A_4 \xrightarrow{a_2} A_1 A_2 B_3 A_4 \xrightarrow{a_3} A_1 A_2 A_3 B_4 \xrightarrow{a_4} \sigma \qquad (1)$$

and

$$B_1 C_2 A_3 A_4 \xrightarrow{b_1} A_1 C_2 A_3 A_4 \xrightarrow{b_2} A_1 A_2 C_3 A_4 \xrightarrow{a_3} A_1 A_2 A_3 C_4 \xrightarrow{a} \sigma \qquad (2)$$

Their nodes are justified by place invariants $A_i + B_i + C_i = 1$ for $i = 1, \ldots, 4$. We leave the details as an exercise to the reader.

As a consequence, each concurrent run of the asynchronous stack has a linearization that consists in a sequence of rounds, each of which describes either an entire push-wave or an entire pop-wave of the stack.

55.3   Rounds of the crosstalk algorithm

Figure 55.2 recalls the crosstalk algorithm $\Sigma_{12.5}$ with renamed places. This algorithm operates in three rounds, as already discussed in Sect. 12. Here we are going to prove that the initial state AE is in fact a ground state. The following place invariants of $\Sigma_{55.2}$ will be used:

inv1: $A + B + G = 1,$
inv2: $E + K + F = 1,$
inv3: $A + C + D + J + M = 1,$
inv4: $D + E + H + J + L = 1.$

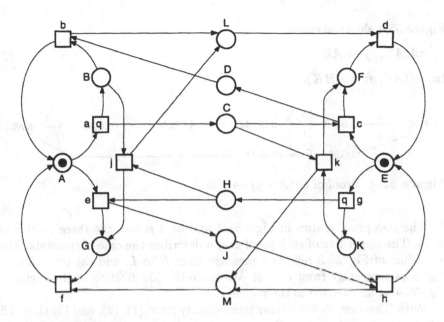

Figure 55.2. Renamed crosstalk $\Sigma_{12.5}$

First we observe that

$\{a, g\}$ is a change set of AE: (1)

A prevents c by inv3 and E prevents e by inv4. Figure 55.3 shows

$\text{eff}(AE, a) \hookrightarrow AE$ (2)

(as $\text{eff}(AE, a) = BCE$). The nodes of Fig. 55.3 are justified as follows:

node 1: context B, and E prevents k by inv2;
node 2: context BH, K prevents c by inv2, and C prevents h by inv3;
node 3: context FM, B prevents e by inv1, and H prevents b by inv4;
node 4: context GM;
node 5: context E;
node 6: context F, D prevents j by inv4;
node 7: context A.

Figure 55.3. Proof of $\text{eff}(AE, a) \hookrightarrow AE$

Figure 55.4 shows likewise

$$\mathrm{eff}(AE, g) \hookrightarrow AE \tag{3}$$

(as $\mathrm{eff}(AE, g) = AHK$).

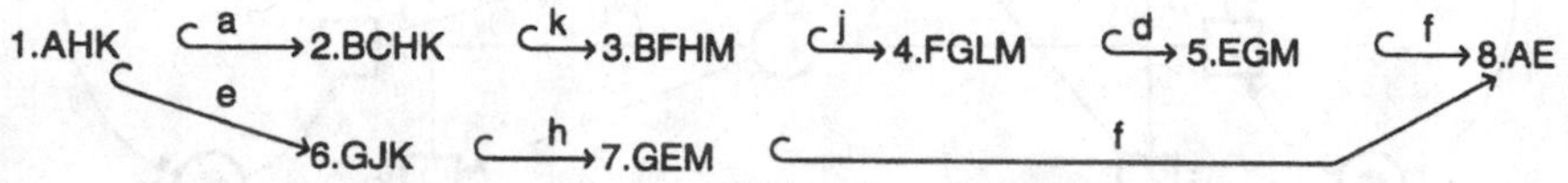

1.AHK $\xrightarrow{\;a\;}$ 2.BCHK $\xrightarrow{\;k\;}$ 3.BFHM $\xrightarrow{\;j\;}$ 4.FGLM $\xrightarrow{\;d\;}$ 5.EGM $\xrightarrow{\;f\;}$ 8.AE

$\xrightarrow{\;e\;}$ 6.GJK $\xrightarrow{\;h\;}$ 7.GEM $\quad f$

Figure 55.4. Proof of $\mathrm{eff}(AE, g) \hookrightarrow AE$

The two proof graphs in Figs. 55.3 and 55.4 reflect the three rounds of $\Sigma_{55.2}$: The upper line of each proof graph describes the case of crosstalk. The lower line of Fig. 55.3 reflects a message from R to L, and the lower line of Fig. 55.4 a message from L to R. We leave the justification of the nodes of Fig. 55.4 as an exercise to the reader.

With Theorem 53.2 it follows immediately from (1), (2), and (3) that AE is in fact a ground state of $\Sigma_{55.2}$.

56   Ground Formulas of Mutex Algorithms

56.1   Construction of ground formulas

A mutex algorithm can never reach a state that is entirely symmetrical with respect to the two sites involved. If it could, then both sites could continue to always act symmetrically and thus would never reach a state with one site critical and the other site not critical. In fact, each algorithm of Sect. 11 is either asymmetrical by structure (such as the synchronized algorithm, Owicki/Lamport's algorithm, and the asymmetrical algorithm) or is asymmetrical by initial state (such as the token passing algorithm, Peterson's algorithm, and Dekker's algorithm). The contentious, the alternating, and the state-testing algorithms are symmetrical both in structure and initial state, but have been ruled out as acceptable mutex algorithms.

Asymmetry of the initial state is inherited to each reachable state, and thus the token-passing, Peterson's, and Dekker's algorithm don't have a ground state. But the local *quiet* states of the two sites l and r constitute a ground formula

$$quiet_l \wedge quiet_r \tag{1}$$

in each mutex algorithm. The initial state of the three structurally asymmetrical algorithms are even ground states.

Proof of (1) can again be based on Theorem 53.2. In each mutex algorithm Σ, the initial state a_Σ enables two actions which constitute a change set of a_Σ, and lead to $pend_l$ and $pend_r$ (or similarly denoted local states), respectively. So we have to show

$$pend_l \wedge quiet_r \hookrightarrow a_\Sigma, \quad \text{and} \quad quiet_l \wedge pend_r \hookrightarrow a_\Sigma. \tag{2}$$

Evolution of Σ gives $pend_l \hookrightarrow crit_l$ and $pend_r \hookrightarrow crit_r$. A proof of evolution of various algorithms, as given in Sect. 49, can easily be modified to proofs of

$$\begin{aligned} &pend_l \wedge quiet_r \hookrightarrow crit_l \wedge quiet_r, \quad \text{and} \\ &quiet_l \wedge pend_r \hookrightarrow quiet_l \wedge crit_r. \end{aligned} \tag{3}$$

Furthermore,

$$crit_l \wedge quiet_r \hookrightarrow a_\Sigma, \quad \text{and} \quad quiet_l \wedge crit_r \hookrightarrow a_\Sigma. \tag{4}$$

can immediately be picked up by Corollary 51.2 (with context $quiet_r$ and $quiet_l$, respectively).

Hence (2) follows from (3) and (4), and the transitivity of the causes operator $\hookrightarrow$.

All this does not yet apply to the round-based mutex algorithm $\Sigma_{13.6}$, because its evolution remains to be proven.

56.2 A ground formula of the round-based mutex algorithm

Figure 56.1 recalls the round-based mutex algorithm $\Sigma_{13.6}$, with renamed places. We will show

$$AE \text{ is a ground formula of } \Sigma_{56.1} \tag{1}$$

Proof of (1) employs the following place invariants of $\Sigma_{56.1}$:

inv1: $A + B + G = 1,$
inv2: $E + F + K = 1,$
inv3: $A + C + D + J + M + R = 1,$
inv4: $D + E + H + J + L + N = 1.$

First we observe that

$$\{a, g\} \text{ is a change set of } AE \tag{2}$$

as A prevents c by inv3 and E prevents e by inv4. As a technicality, Fig. 56.2 shows $BCHK \hookrightarrow AE$. The nodes of Fig. 56.2 are justified as follows:

node 1: context CK, C prevents b by inv3 and B prevents e by inv1;
node 2: context CGK;
node 3: context G, K prevents c by inv2, K prevents d by inv2 and C prevents h by inv3;
node 4: context EG;
node 5: context E.

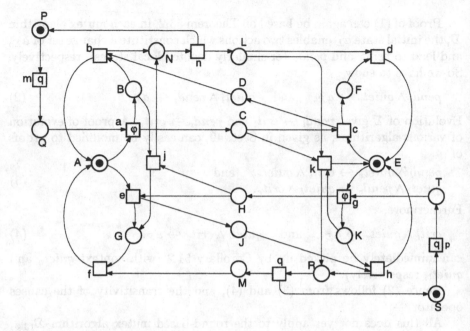

Figure 56.1. Renamed round-based mutex algorithm $\Sigma_{13.6}$

$$1.\text{BCHK} \xrightarrow{\;\;j\;\;} 2.\text{CGKN} \xrightarrow{\;\;n\;\;} 3.\text{CGKL} \xrightarrow{\;\;k\;\;} 4.\text{EGR} \xrightarrow{\;\;q\;\;} 5.\text{EGM} \xrightarrow{\;\;f\;\;} 6.\text{AE}$$

Figure 56.2. $\Sigma_{56.1} \models BCHK \hookrightarrow AE$

Now Fig. 56.3 shows

$$\text{eff}(AE, a) \hookrightarrow AE \tag{3}$$

$$1.\text{BCE} \xrightarrow{\;\;c\;\;} 2.\text{BDF} \xrightarrow{\;\;b\;\;} 3.\text{AFN} \xrightarrow{\;\;h\;\;} 4.\text{AFL} \xrightarrow{\;\;d\;\;} 6.\text{AE}$$
$$\xrightarrow{\;a\,?\;} 5.\text{BCHK} \xleftarrow{\qquad\text{Fig.56.2}\qquad}$$

Figure 56.3. $\Sigma_{56.1} \models \text{eff}(AE, a) \hookrightarrow AE$

The nodes of Fig. 56.3 are justified as follows:

node 1: context B, and E prevents k by inv2;
node 2: context F, and D prevents j by inv4;
node 3: context AF;
node 4: context A, and F prevents k by inv2;
node 5: Figure 56.2.

Finally, Fig. 56.4 shows

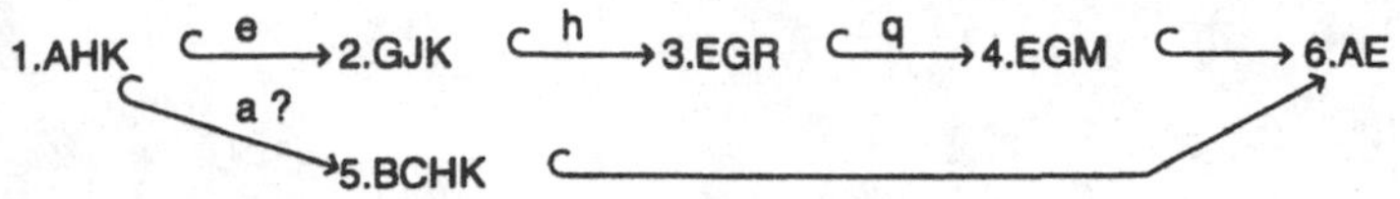

1.AHK $\xrightarrow{\ e\ }$ 2.GJK $\xrightarrow{\ h\ }$ 3.EGR $\xrightarrow{\ q\ }$ 4.EGM $\longrightarrow$ 6.AE

a ?

5.BCHK

Figure 56.4. $\Sigma_{56.1} \models \mathrm{eff}(AE, g) \hookrightarrow AE$

$$\mathrm{eff}(AE, g) \hookrightarrow AE \tag{4}$$

The nodes of Fig. 56.4 are justified as follows:

- node 1: context K, and A prevents j by inv1;
- node 2: context G, and J prevents k by inv3;
- node 3: context EG;
- node 4: context E;
- node 5: Figure 56.2.

Altogether, (2), (3), and (4) imply (1) by Theorem 53.2.

56.3 Evolution of the round-based mutex algorithm

Due to the asymmetry of the round-based mutex algorithm we have to show both $\Sigma_{13.6} \models pend1_l \mapsto crit_l$ and $\Sigma_{13.6} \models pend1_r \mapsto crit_r$. In the representation of Fig. 56.1 this reads

$$\Sigma_{56.1} \models Q \mapsto N, \quad \text{and} \tag{1}$$

$$\Sigma_{56.1} \models T \mapsto R \tag{2}$$

Proof of these properties employs the ground formula AE: First we observe that the ground formula AE implies true $\hookrightarrow A$ (by Def. 53.1), hence $Q \hookrightarrow A$ (by propositional logic), hence $Q \mapsto A$ (by Lemma 50.4(ii)). This in turn implies

$$\Sigma_{56.1} \models Q \mapsto B \tag{3}$$

by Theorem 48.2. Furthermore, $\Sigma_{56.1} \models B \to \neg A$ (by inv1) and $\{b, j\}$ is a change set of $\{B\}$. Hence $\Sigma_{48.1} \models B \hookrightarrow N$ (by Theorem 53.3), hence

$$\Sigma_{56.1} \models B \mapsto N \tag{4}$$

by Lemma 50.4(ii). Thus (1) follows from (3) and (4) with Lemma 44.3(iii). Likewise, one shows $\Sigma_{56.1} \models T \mapsto K$ and $\Sigma_{56.1} \models K \mapsto R$, which implies (2).

Part D
Analysis of Advanced System Models

In analogy to the analysis of elementary system nets as described in Part C, state properties and progress properties will be considered separately for system nets, too.

X. State Properties of System Nets

State properties have been defined for elementary system nets as propositional combinations on a system's local states. Particularly important state properties have been derived from valid equations and inequalities formed $n_1 \cdot p_1 + \cdots + n_k \cdot p_k = m$ and $n_1 \cdot p_1 + \cdots + n_k \cdot p_k \geq m$, respectively. Each integer n_i provides a *weight* for p_i. This kind of equation or inequality *holds* at a state s if valuation of variables p_i by the integer $s(p_i)$ solves the equation or inequality.

For advanced system nets, any domain D may provide weights $f(p_i) \in D$ for a place p_i. The function f must be applicable to the contents $s(p_i)$ of p_i at any reachable state, s. In fact, this approach is followed for system nets in the sequel. As an example, in the term represented system net

sort	dom
const	u : dom
fct	f, g : dom → dom
var	x : dom

$$\tag{1}$$

the number of tokens remains invariant, provided the tokens on A are counted twice: For each reachable state s, $2 \cdot |s(A)| + |s(B)| = 2$. As a shorthand this is represented by the symbolic equation

$$2|A| + |B| = 2. \tag{2}$$

A more informative state property is gained by weight functions f and g, canonically extended to sets and coincidentally applied to the token load of A. In fact, at each reachable state s, $f(s(A)) \cup g(s(A)) \cup s(B) = \{f(u), g(u)\}$. As a matter of convention and unification, this will be expressed by the symbolic equation

$$f(A) + g(A) + B = f(u) + g(u). \tag{3}$$

The unifying formal background for both (2) and (3) are *multisets* of items, in which an item may occur more than once. Multisets and linear functions on multisets provide means to construct equations, inequalities, place invariants, and initialized traps for system nets. All those concepts are generalizations of the corresponding concepts for es-nets given in Chap. VII.

57   First-Order State Properties

Formulas to represent properties of states of advanced system nets will be employed, by analogy to formulas to represent properties of states of elementary system nets, as introduced in Sect. 38. Terms as introduced in Sect. 19 (there used as arc inscriptions) will serve in a first-order logic, with places of system nets as predicate symbols (by analogy to Sect. 39.1, where places of elementary system nets served as propositional variables).

We start with the syntax of formulas over a structure $\mathcal{A}$.

57.1 Definition. *Let $\mathcal{A}$ be a structure, let X be a set of $\mathcal{A}$-sorted variables, and let P be any set of symbols. Then the set $\mathcal{F}(\mathcal{A}, X, P)$ of state formulas over $\mathcal{A}$, X, and P is the smallest set of symbol chains such that for all $t \in T_{\mathcal{A}}(X)$ and all $p, q \in P$,*

  i. *$p.t$, $p = t$, and $p \subseteq q \in \mathcal{F}(\mathcal{A}, X, P)$*
  ii. *if $f, g \in \mathcal{F}(\mathcal{A}, X, P)$ then $f \wedge g \in \mathcal{F}(\mathcal{A}, X, P)$ and $\neg f \in \mathcal{F}(\mathcal{A}, X, P)$.*

The following notations will be used, by analogy to Sect. 38.2:

57.2 Notations. *In the sequel we employ the conventional propositional symbols $\vee$ and $\rightarrow$, and for any set $Q = \{q_1, \ldots, q_n\}$ the shorthands $\bigvee Q$ for $q_1 \vee \ldots \vee q_n$, and $\bigwedge Q$ or just Q for $q_1 \wedge \ldots \wedge q_n$. Furthermore, we write $\mathcal{A}.u_1, \ldots, u_n$ as a shorthand for $\mathcal{A}.u_1 \wedge \ldots \wedge \mathcal{A}.u_n$.*

Each advanced system net Σ is assigned its set of state formulas. Those formulas are constructed from the structure of Σ, with the places of Σ serving as predicate symbols. The token load $s(p)$ of place p at a state s, as well as the inscriptions in $\overline{f}$ of an arc f, are terms that may occur in state formulas.

57.3 Definition. *Let $\mathcal{A}$ be a structure, let X be an $\mathcal{A}$-sorted set of variables, and let Σ be a net, term-inscribed over $\mathcal{A}$ and X.*

  i. *Each $f \in \mathcal{F}(\mathcal{A}, X, P_\Sigma)$ is a state formula of Σ.*
  ii. *For each state s of Σ, the state formula $\hat{s}$ of Σ is defined by $\hat{s} := \bigwedge_{p \in P_\Sigma} s(p)$.*

Such formulas are interpreted as follows:

57.4 Definition. *Let Σ be an es-net, let f be a state formula of Σ, let v be an argument for its variables, and let s be a state of Σ.*

  i. *$s \models f(v)$ ("a is an $f(v)$-state") is inductively defined over the structure of f. To this end, let $u \in T_{\mathcal{A}}(X)$, $p, q \in P_\Sigma$ and $g, h \in \mathcal{F}(\mathcal{A}, X, P)$.*
    - *$s \models p.t(v)$ iff $\mathrm{setval}^u(v) \subseteq s(p)$, and*
      $s \models (p = t)(v)$ iff $\mathrm{setval}^u(v) = s(p)$.
    - *$s \models p \subseteq q$ iff $s(p) \subseteq s(q)$.*
    - *$s \models g \wedge h$ iff $s \models g$ and $s \models h$.*
    - *$s \models \neg g$ iff not $s \models g$.*

ii. $s \models f$ iff, for all arguments u of X, $s \models f(u)$.
iii. $\Sigma \models f$ iff, for all reachable states s of Σ, $s \models f$.

Apparently, for each state a, $a \models \hat{a}$.

58   Multisets and Linear Functions

State properties can frequently be proven by means of equations and inequalities, which in turn can be derived from the static structure of a given system
net, by analogy to equations and inequalities of es-nets. Each place of the
net will serve as a variable, ranging over the subsets of the places' domains.
Terms will employ *linear extensions* of functions of the underlying algebra.

Each structure A canonically induces *multisets* of its carrier sets and linear
extensions of its functions. Intuitively, a multiset B over a set A assigns to
each $a \in A$ a multiplicity of occurrences of a. As a special case, a conventional
subset of a sticks to the multiplicities 0 and 1. For technical convenience we
allow negative multiplicities, too, But *proper* multisets have no negative entry.

58.1 Definition. *Let A be a set.*

i. Any function $M : A \rightarrow \mathbb{Z}$ is called a multiset *over A. Let $A^{\mathfrak{M}}$ denote the
set of all multisets over A.*
*ii. Let $M \in A^{\mathfrak{M}}$ and $z \in ZZ$. Then $zM \in A^{\mathfrak{M}}$ is defined for each $a \in A$ by
$zM(a) := z \cdot M(a)$.*
*iii. Let $L, M \in A^{\mathfrak{M}}$. Then $L + M \in A^{\mathfrak{M}}$ is defined for each $a \in A$ by
$(L + M)(a) := L(a) + M(a)$.*
iv. A multiset $M \in A^{\mathfrak{M}}$ is proper *iff $M(a) \geq 0$ for all $a \in A$.*

Sets can be embedded canonically into multisets, and some operations on
sets conditionally correspond to operations on multisets:

58.2 Definition. *Let A be a set, let $a \in A$ and $B \subseteq A$. If A is obvious from
the context, a^m and B^m denote multisets over A, defined by $a^m(x) = 1$ if
$x = a$ and $a^m(x) = 0$ otherwise; and $B^m(x) = 1$ if $x \in B$ and $B^m(x) = 0$,
otherwise.*

Union and difference of sets correspond to addition and subtraction of the
corresponding multisets, given some additional assumptions:

58.3 Lemma. *Let A be a set and let $B, C \subseteq A$.*

i. $(B \cup C)^m = B^m + C^m$, provided $B \cap C = \emptyset$ and
ii. $(B \setminus C)^m = B^m - C^m$, provided $C \subseteq B$.

58.4 Notations.

i. By abuse of notation we usually write just A instead of A^m.

ii. Addition $B + C$ and subtraction $B - C$ are written for ordinary sets B and C only if $B \cap C = \emptyset$ and $C \subseteq B$, respectively. Particularly, for $a \in A$ and $B \subseteq A$, $B - a$ is written only if $a \in B$.

There is a canonically defined scalar product and a sum of functions over multisets:

58.5 Definition. *Let A and B be sets:*

i. Any function $\varphi : A^{\mathfrak{M}} \to B^{\mathfrak{M}}$ is called a multiset function from A to B.

ii. Let $\varphi : A^{\mathfrak{M}} \to B^{\mathfrak{M}}$ be a multiset function and let $z \in \mathbb{Z}$. Then $z\varphi : A^{\mathfrak{M}} \to B^{\mathfrak{M}}$ is defined for each $M \in A^{\mathfrak{M}}$ by $z\varphi(M) := z \cdot (\varphi(M))$.

iii. Let $\varphi, \psi : A^{\mathfrak{M}} \to B^{\mathfrak{M}}$ be two multiset functions. Then $\varphi + \psi : A^{\mathfrak{M}} \to B^{\mathfrak{M}}$ is defined for each $M \in A^{\mathfrak{M}}$ by $(\varphi + \psi)(M) := \varphi(M) + \psi(M)$.

iv. $\mathbb{O}_{AB}$ denotes the zero-valuating multiset function from A to B, i.e., $\mathbb{O}_{AB}(M) = \mathbb{O}_B$ for each $M \in A^{\mathfrak{M}}$. The index AB is skipped whenever it can be assumed from the context.

Each function $f : A \to B$ and each set-valued function $g : A \to B^{\mathfrak{M}}$ of a structure $\mathcal{A}$ can be extended canonically to a multiset function $g : A^{\mathfrak{M}} \to B^{\mathfrak{M}}$:

58.6 Definition. *Let A and B be sets and let $f : A \to B$ or $f : A \to B^{\mathfrak{M}}$ be a function. Then the multiset function $\hat{f} : A^{\mathfrak{M}} \to B^{\mathfrak{M}}$ is defined for each $M \in A^{\mathfrak{M}}$ and each $b \in B$ by $\hat{f}(m)(b) = \Sigma_{a \in f^{-1}(b)} M(a)$.*

By abuse of notation we write f instead of $\hat{f}$ whenever the context excludes confusion. The induced functions $\hat{f}$ are *linear*:

58.7 Lemma. *Let A and B be sets, let $f : A \to B$ be a function, let $L, M \in \mathfrak{M}(A)$, and let $z \in \mathbb{Z}$. Then for the multiset extension of f, $\hat{f}(L + M) = \hat{f}(L) + \hat{f}(M)$, and $\hat{f}(z \cdot M) = z \cdot \hat{f}(M)$.*

Proof. Let $b \in B$ and let $C := f^{-1}(b)$.

i. $\hat{f}(L + M)(b) = \Sigma_{a \in C}(L + M)(a) = \Sigma_{a \in C}(L)(a) + \Sigma_{a \in C}(M)(a) = \hat{f}(L)(a) + \hat{f}(M)(a) = (\hat{f}(L) + \hat{f}(M))(a)$.

ii. $\hat{f}(z \cdot M)(b) = \Sigma_{a \in C}(z \cdot M)(a) = \Sigma_{a \in C} z \cdot M(a) = z \cdot \Sigma_{a \in C} M(a) = z \cdot \hat{f}(M)$. $\square$

59   Place Weights, System Equations, and System Inequalities

State properties are essentially based on weighted sets of tokens, formally given by multiset valued mappings on the places' domains.

59.1 Definition. *Let Σ be a system net over a universe A, let $p \in P_\Sigma$, and let B be any multiset. Then a mapping $I : A_p \to B$ is a place weight of p. I is natural if $B = \mathbf{N}$.*

Place weights are frequently extended to set-valued arguments and then applied to the token load $s(p)$ of the token at place p in a global state, s. In this case, a multiset $I(s(p))$ is called a *weighted token load of p*.

Place weights can be used to describe invariant properties of system nets by help of equations that hold in all reachable states:

59.2 Definition. *Let Σ be a system net over a universe A, let B be any multiset and let $P = \{p_1, \ldots, p_n\} \subseteq P_\Sigma$. For $j = 1, \ldots, k$, let $I^j : A_{p_j} \to B$ be a place weight of p_j.*

i. *$\{I^1, \ldots, I^k\}$ is a Σ-invariance with value B if for each reachable state s of Σ,*
$$I^1(s(p_1)) + \cdots + I^k(s(p_k)) = B.$$

ii. *A Σ-invariance $\{I^1, \ldots, I^k\}$ is frequently written as a symbolic equation*
$$I^1(p_1) + \cdots + I^k(p_k) = B$$
and this equation is said to hold *in Σ.*

In a Σ-equation $I^1(p_1) + \cdots + I^k(p_k) = B$, the value of B is apparently equal to $I^1(s_\Sigma(p_1)) + \cdots + I^k(s_\Sigma(p_k))$, with s_Σ the initial state of Σ.

As a technical example, in the term inscribed representation of a system net Σ,

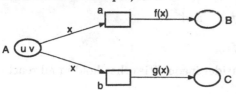

sort	dom
const	u, v : dom
fct	f, g : dom $\to$ dom
var	x : dom

(1)

let $\{u, v\}$ be the domain of both A and B, and for $x \in \{a, b\}$ let $I^A(x) = f(x) + g(x)$ and $I^B(x) = x$. Then $\{I^A, I^B\}$ is a Σ-invariance with value $U = f(u) + g(u) + f(v) + g(v)$, symbolically written

$$f(A) + g(A) + B = U. \tag{2}$$

One of the reachable states is s, with $s(A) = u$ and $s(B) = f(v) + g(v)$. Then in fact $I^A(s(A)) + I^B(s(B)) = I^A(u) + I^B(f(v)) + I^B(g(v)) = U$.

Intuitively formulated, according to this invariance, the element u is at A, or both $f(u)$ and $g(u)$ are at B. The corresponding property for v holds accordingly in Σ.

As a further example, in $\Sigma =$

sort	dom
const	u, v : dom
fct	f, g, f^{-1}, g^{-1} : dom $\to$ dom
var	x : dom

$f^{-1}(f(x)) = x$
$g^{-1}(g(x)) = x$

(3)

let again $\{u, v\}$ be the domain of all places A, B, and C, and for $x \in \{u, v\}$ let $I^A(x) = x$, $I^B(x) = f^{-1}(x)$ and $I^C(x) = g^{-1}(x)$. Then $\{I^A, I^B, I^C\}$ is a Σ-invariance with value $u+v$, symbolically written $A+f^{-1}(B)+g^{-1}(C) = u+v$. One of the reachable states is s, with $s(A) = u$, $s(B) = f(v)$ and $s(C) = \emptyset$. Then in fact $I^A(s(A)) + I^B(s(B)) + I^C(s(C)) = I^A(u) + I^B(f(v)) + I^C(\emptyset) = u + f^{-1}(f(v)) = u + v$.

As a final technical example, in $\Sigma =$

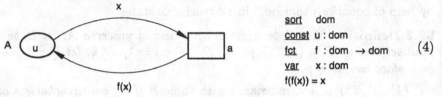

sort   dom
const u : dom
fct   f : dom $\rightarrow$ dom   (4)
var   x : dom
f(f(x)) = x

let $U = \{u\}$ and $I^A(x) = x + f(x)$. Then $\{I^A\}$ is a Σ-invariance with value $u + f(u)$, symbolically written

$$A + f(A) = u + f(u).$$

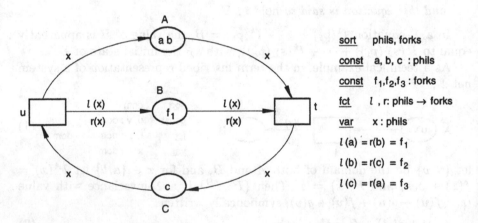

sort   phils, forks
const a, b, c : phils
const f_1, f_2, f_3 : forks
fct l , r: phils $\rightarrow$ forks
var   x : phils
$l(a) = r(b) = f_1$
$l(b) = r(c) = f_2$
$l(c) = r(a) = f_3$

Figure 59.1. Renamed philosophers system $\Sigma_{15.10}$

A more realistic example is the philosophers system $\Sigma_{15.10}$, redrawn in Fig. 59.1. This system has three interesting equations:

$$A + C = a + b + c,$$

$$B + l(C) + r(C) = f_1 + f_2 + f_3, \quad \text{and}$$

$$r(A) + l(A) - B = f_1 + f_2 + f_3.$$

In analogy to Sect. 39 we also consider inequalities that hold in all reachable states:

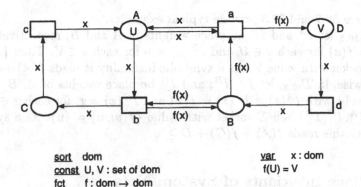

sort	dom	var	x : dom
const	U, V : set of dom		f(U) = V
fct	f : dom → dom		

Figure 59.2. $f(A) + B \geq V$ is a valid inequality

59.3 Definition. *Let Σ be a system net over a universe A, let B be any multiset, and let $P = \{p_1, \ldots, p_k\} \subseteq P_\Sigma$. For $j = 1, \ldots, k$ let $I^j : A_{p_j} \to B$ be a place weight of p.*

$\{I^1, \ldots, I^k\}$ yields a Σ-socket with value B if for each reachable state s of Σ,

$$I^1(s(p_1)) + \cdots + I^k(s(p_k)) \geq B.$$

A Σ-socket $\{I^1, \ldots, I^k\}$ is frequently written as a symbolic inequality

$$I^1(p_1) + \cdots + I^k(p_k) \geq B,$$

and this inequality is said to hold in Σ.

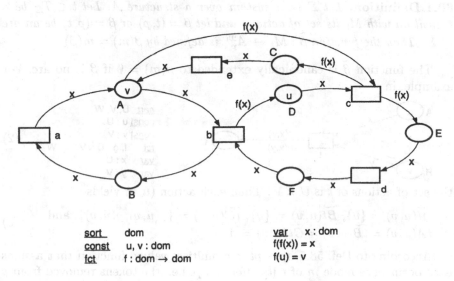

sort	dom	var	x : dom
const	u, v : dom		f(f(x)) = x
fct	f : dom → dom		f(u) = v

Figure 59.3. $f(A) + f(C) + D \geq u$ is a valid inequality

Figures 59.2 and 59.3 provide typical examples.

In $\Sigma_{59.2}$ let I^A and I^B be place weights of A and B, respectively, with $I^A(x) = f(x)$ for each $x \in U$ and $I^B(y) = y$ for each $y \in V$. Then $\{I^A, I^B\}$ is a Σ-socket with value V. As a symbolic inequality it reads $f(A) + B \geq V$.

Likewise, in $\Sigma_{59.3}$, let I^A, I^B, and I^C be place weights of A, B, and C, respectively, with $I^A(x) = I^C(x) = f(x)$ and $I^D(x) = x$, for each $x \in \{u, v\}$. Then $\{I^A, I^C, I^D\}$ is a Σ-socket with value $\{f(v), u\} = \{u\}$. As a symbolic inequality this reads $f(A) + f(C) + D \geq u$.

60 Place Invariants of System Nets

We are now seeking a technique to prove Σ-invariances without explicitly visiting all reachable states. To this end we construct *place invariants* for system nets, in analogy to place invariants of es-nets: A set of place weights is a place invariant if each occurrence mode m of each transition t yields a balanced weighted effect to the places involved, i.e., the weighted set of removed tokens is equal to the weighted set of augmented tokens; formally, for place weights $I^1, \ldots, I^k$ of places $p_1, \ldots, p_k$,

$$I^1(m(t, p_1)) + \cdots + I^k(m(t, p_k)) = I^1(m(p_1, t)) + \cdots + I^k(m(p_k, t)). \quad (1)$$

A more concise representation of (1) is gained by a slightly different perspective on transitions and their actions: Each arc $\beta = (p, t)$ or $\beta = (t, p)$ defines a mapping $\widetilde{\beta}$ that assigns each action m of t the corresponding subset $m(\beta)$ of A_p. Furthermore, this subset is canonically conceived as a multiset, i.e., an element of $A_p^{\mathfrak{M}}$:

60.1 Definition. *Let Σ be a system over a structure $\mathcal{A}$. Let $t \in T_\Sigma$ be a transition with M_t its set of actions and let $\beta = (t, p)$ or $\beta = (p, t)$ be an arc of Σ. Then the function $\widetilde{\beta} : M_t \to A_p^{\mathfrak{M}}$ is defined by $\widetilde{\beta}(m) = m(\beta)$.*

The function $\widetilde{\beta}$ is canonically extended to $\widetilde{\beta}(m) = \emptyset$ if β is no arc. For example, in

the set of actions of t is $U \times V$. Then each action (u, v) yields

$$\widetilde{At}(u, v) = \{u\}, \ \widetilde{Bt}(u, v) = \{v\}, \ \widetilde{tC}(u, v) = \{f(u, v), g(u, v)\}, \text{ and} \quad (3)$$
$$\widetilde{tA}(u, v) = \widetilde{tB}(u, v) = \widetilde{Ct}(u, v) = \emptyset.$$

According to Def. 58.5, $\widetilde{tp} - \widetilde{pt}$ is a multiset valued function that assigns each occurrence mode m of t its effect on p, i.e., the tokens removed from p or augmented to p upon t's occurrence in mode m.

Each place weight $I^p : A_p \to B$ of a place p can canonically be extended to the set valued arguments $I^p : A_p^{\mathfrak{M}} \to B^{\mathfrak{M}}$, by Def. 58.6. This function in turn can be composed with $\widetilde{tp} - \widetilde{pt}$, yielding a function $I^p \circ (\widetilde{tp} - \widetilde{pt}) : M_t \to B^{\mathfrak{M}}$.

A set of place weights is a *place invariant* if the sum of weighted effects of all involved places reduces to the zero function $\mathbb{O}$. The *value* of a place invariant is derived from the net's initial state:

60.2 Definition. *Let Σ be a system net and let $p_1, \ldots, p_k \in P_\Sigma$. For $j = 1, \ldots, k$ let I^j be a place weight of p_j. Then $I = \{I^1, \ldots, I^k\}$ is a place invariant of Σ if for each transition $t \in T_\Sigma$,*

$$I^1 \circ (\widetilde{tp_1} - \widetilde{p_1 t}) + \cdots + I^k \circ (\widetilde{tp_k} - \widetilde{p_k t}) = \mathbb{O}.$$

The multiset $I^1(s_\Sigma(p_1)) + \cdots + I^k(s_\Sigma(p_k))$ is the value *of I.*

As an example, for the net (2) let I^A, I^B, and I^C be place weights for A, B, and C, respectively, with $I^A(x) = f(x)$ for each $x \in U$, $I^B(y) = g(y)$ for each $y \in V$, and $I^C(z) = z$ for each $z \in W$. Then the set $\{I^A, I^B, I^C\}$ is a place invariant of (2): With (3) follows $I^A \circ (\widetilde{tA} - \widetilde{At}) + I^B \circ (\widetilde{tB} - \widetilde{Bt}) + I^C \circ (\widetilde{tC} - \widetilde{Ct}) = f \circ (\mathbb{O} - \widetilde{At}) + g \circ (\mathbb{O} - Bt) + \widetilde{tC} - \mathbb{O} = -f \circ \widetilde{At} - g \circ \widetilde{Bt} + \widetilde{tC}$. Then for all $(u, v) \in U \times V$, again with (2), $(-f \circ \widetilde{At} - g \circ \widetilde{Bt} + \widetilde{tC})(u, v) = -f(\widetilde{At}(u, v)) - g(\widetilde{Bt}(u, v)) + \widetilde{tC}(u, v) = -f(u, v) - g(u, v) + f(u, v) + g(u, v) = \mathbb{O}$. The value of this place invariant is $I^A(u) + I^B(v) + I^C(\mathbb{O}) = f(u) + g(v)$.

A place invariant provides in fact a valid Σ-equation:

60.3 Theorem. *Let Σ be a system net, let $p_1, \ldots, p_k \in P_\Sigma$, and for $j = 1, \ldots, k$, let I^j be a place weight of Σ. Let $\{I^1, \ldots, I^k\}$ be a place invariant of Σ and let U be its value. Then the equation*

$$I^1(p_1) + \cdots + I^k(p_k) = U$$

holds in Σ.

Proof.   i. Let $r \xrightarrow{t,m} s$ be a step of Σ. Then for each $p \in P_\Sigma$, $s(p) = r(p) + m(t, p) - m(p, t)$, by Proposition 16.4. Then

$$\Sigma_{j=1}^k I^j(s(p_j)) = \Sigma_{j=1}^k I^j(r(p_j) + m(t, p_j) - m(p_j, t))$$

$$= \Sigma_{j=1}^k I^j(r(p_j)) + \Sigma_{j=1}^k I^j(m(t, p_j) - m(p_j, t)) \qquad \text{by Def. 58.1}$$

$$= \Sigma_{j=1}^k I^j(r(p_j)) + \Sigma_{j=1}^k I^j((\widetilde{t, p_j})(m) - (\widetilde{p_j, t})(m)) \qquad \text{by Def. 19.2}$$

$$= \Sigma_{j=1}^k I^j(r(p_j)) + \Sigma_{j=1}^k I^j((\widetilde{t, p_j}) - (\widetilde{p_j, t}))(m) \qquad \text{by Lemma 58.7}$$

$$= \Sigma I^j(r(p_j)) + \mathbb{O}(m) = \Sigma I^j(r(p_j))$$

ii. Now let s be a reachable state of Σ. Then there exists an interleaved run of Σ formed $s_0 \xrightarrow{t_1,m_1} s_1 \xrightarrow{t_2,m_2} \ldots \xrightarrow{t_l,m_l} s_l$ with $s_l = s$. Then $\Sigma_{j=1}^{k} I^j(s_0(p_j)) = U$, by Def. 60.2. Then for each $i = 1, \ldots, l$, $\Sigma_{j\leq1}^{k} I^j(s_i(p_j)) = U$, by i. This yields the proposition for $i = l$. $\square$

Place invariants can be mimicked symbolically in term-inscribed representations of system nets. To this end, the functions $\widetilde{tp}$, $\widetilde{pt}$, $\widetilde{tp} - \widetilde{pt}$, and I^p will be represented symbolically. The composition $I^p \circ (\widetilde{tp} - \widetilde{pt})$ of functions I^p and $(\widetilde{tp} - \widetilde{pt})$ then is symbolically executable as substitution of terms.

Definition 19.1 assigns each arc $\beta = (t,p)$ or $\beta = (p,t)$ of a term-inscribed net Σ a set $\overline{\beta} \subseteq T_{A_p}(X_t)$ of A_p-terms over X_t. For each $u \in \overline{\beta}$, val^u (as defined in Def. 18.5) is a mapping from M_t to A_p. This mapping can be extended canonically to $val^u : M_t \to A_p^{\mathfrak{M}}$. Mappings of this kind can be summed up, giving rise to the mapping $\widetilde{\beta} : M_t \to A_p^{\mathfrak{M}}$ of Def. 60.1, defined by $\widetilde{\beta}(m) := val^{u_1}(m) + \cdots + val^{u_k}(m)$, with $X_t = \{u_1, \ldots, u_k\}$. Hence $\widetilde{\beta}$ can be represented symbolically as

$$\widetilde{\beta} = u_1 + \cdots + u_k \tag{1}$$

in this case.

The multiset extension $I^p : A_p^{\mathfrak{M}} \to B$ of a place weight $I : A_p \to B$ can be represented as a term with one variable, ranging over $A_p^{\mathfrak{M}}$. For the sake of convenience we always choose the variable p, hence the corresponding term is an element of $T_B(\{p\})$.

The composed function $I^p \circ (\widetilde{tp} - \widetilde{pt}) : M_t \to B$ is now symbolically represented by the multiset term

$$\tau = I^p[\widetilde{tp} - \widetilde{pt}/\mathrm{p}] \tag{2}$$

which is gained from I^p by replacing each occurrence of the variable p in I^p by the term $\widetilde{tp} - \widetilde{pt}$. Hence τ is a term in $T_B(X_t)$, and its valuation val^τ is equal to $I^p \circ (\widetilde{tp} - \widetilde{pt})$.

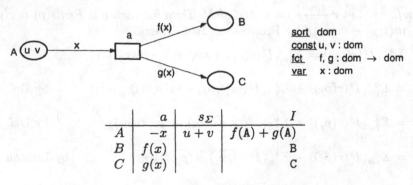

	a	s_Σ	I
A	$-x$	$u + v$	$f(A) + g(A)$
B	$f(x)$		B
C	$g(x)$		C

Figure 60.1. System net with matrix, initial state s_Σ, and a place invariant

The analogy to Sect. 40 continues, as a term-inscribed net Σ is represented as a matrix $\underline{\Sigma}$ with row indices P_Σ, column indices T_Σ, and entries $\underline{\Sigma}(p, t) = \tilde{t}p - \tilde{p}t$. Its initial state s_Σ, as well as each place invariant I, can be represented as a column vector, representing the initial token load $s_\Sigma(p)$ as a variable free ground term of sort A_p and each entry $I(p)$ as the term I^p, introduced above. Moreover, place invariants I can be characterized as solutions of

$$\underline{\Sigma} \cdot I = (\mathbb{O}, \ldots, \mathbb{O}) \tag{3}$$

with $\mathbb{O}$ a symbol for the zero multiset function, as described in Def. 58.5. I then is a vector of place weights, one for each place. The product of a component I^p of I with a matrix entry $\underline{\Sigma}(p, t)$ is the substitution $I^p[\underline{\Sigma}(p, t)/\mathrm{p}]$, addition of terms is the symbolic sum of multiset terms.

As an example, Fig. 60.1 shows a system net together with its matrix and the vector representation of its initial state and a place invariant. Substitution of matrix entries into the components of I yields

$$\begin{aligned}
I^A[\overline{aA} - \overline{Aa}/\mathrm{A}] &= I^A[-x/\mathrm{A}] \\
&= f(\mathrm{A}) + g(\mathrm{A})[-x/\mathrm{A}] \\
&= f(-x) + g(-x) \\
&= -f(x) - g(x),
\end{aligned}$$

$$\begin{aligned}
I^B[\overline{aB} - \overline{Ba}/\mathrm{B}] &= I^B[f(x)/\mathrm{B}] \\
&= \mathrm{B}[f(x)/\mathrm{B}] \\
&= f(x),
\end{aligned}$$

$$\begin{aligned}
I^C[\overline{aC} - \overline{Ca}/\mathrm{C}] &= I^C[g(x)/\mathrm{C}] \\
&= \mathrm{C}[g(x)/\mathrm{C}] \\
&= g(x).
\end{aligned}$$

Figure 60.2 likewise provides the matrix, the initial state, and a place invariant of the net (3) of Sect. 59. Substitution of entries of the first column of the matrix into the components of I yields

$$I^A[\overline{aA} - \overline{Aa}/\text{A}] = I^A[-x/\text{A}]$$
$$= \text{A}[-x/\text{A}]$$
$$= -x,$$

$$I^B[\overline{bB} - \overline{Bb}/\text{B}] = I^B[f(x)/\text{B}]$$
$$= f^{-1}(B)[f(x)/\text{B}]$$
$$= f^{-1}(f(x))$$
$$= x,$$

$$I^C[\text{O}] = \text{O}$$

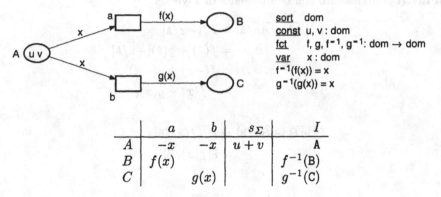

	a	b	s_Σ	I
A	$-x$	$-x$	$u+v$	A
B	$f(x)$			$f^{-1}(B)$
C		$g(x)$		$g^{-1}(C)$

Figure 60.2. Matrix, initial state, and a place invariant to (2) of Sect. 59

As a final technical example, Fig. 60.3 gives matrix, initial state, and a place invariant to (4) of Sect. 59. Substitution of the matrix entry into the invariant yields

$$I^A[\overline{aA} - \overline{Aa}/\text{A}] = I^A[f(x) - x/\text{A}]$$
$$= \text{A} + f(\text{A})[f(x) - x/\text{A}]$$
$$= f(x) - x + f(f(x) - x)$$
$$= f(x) - x + f(f(x)) - f(x)$$
$$= f(x) - x + x - f(x)$$
$$= \text{O}$$

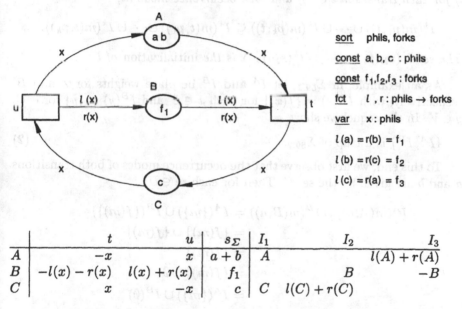

Figure 60.3. Matrix, initial state, and a place invariant to (3) of Sect. 59

To finish this section, Fig. 60.4 shows matrix, initial state, and three place invariants for the philosophers system of Fig. 59.1.

	t	u	s_Σ	I_1	I_2	I_3
A	$-x$	x	$a+b$	A		$l(A)+r(A)$
B	$-l(x)-r(x)$	$l(x)+r(x)$	f_1		B	$-B$
C	x	$-x$	c	C	$l(C)+r(C)$	

Figure 60.4. Matrix, initial state, and three place invariants to $\Sigma_{59.1}$

61   Traps of System Nets

We are now seeking a technique to prove Σ-sockets without visiting all reachable states. To this end we construct *initialized traps* for system nets, in analogy to initialized traps of elementary system nets.

Informally stated, a trap of a system net is a set $\{I^1, \ldots, I^k\}$ of weights of places $p_1, \ldots, p_k$ such that for each element b of a given set B, each transition that removes at least one token with weight b from those places returns at least one token with weight b to those places. This gives rise to an inequality of the form

$$I^1(p_1) + \cdots + I^k(p_k) \geq B. \tag{1}$$

Traps are essentially a matter of plain sets (whereas place invariants are based on multisets). For an arc (p,t) and an occurrence mode m of t, $m(p,t)$ is a plain set according to Def. 16.2. Then $I(m(p,t)) := \{I(u) \mid u \in m(p,t)\}$ is a set, for any place weight I. Construction of traps now goes with set union (not with multiset addition).

61.1 Definition. *Let Σ be a system net and let $p_1, \ldots, p_k \in P_\Sigma$. For $j = 1, \ldots, k$, let I^j be a place weight of p_j. Then $I = \{I^1, \ldots, I^k\}$ is a trap of Σ if for each transition $t \in T_\Sigma$ and each occurrence mode m,*

$$I^1(m(p_1,t)) \cup \cdots \cup I^k(m(p_k,t)) \subseteq I^1(m(t,p_1)) \cup \cdots \cup I^k(m(t,p_k)).$$

The set $I^1(s_\Sigma(p_1)) \cup \cdots \cup I^k(s_\Sigma(p_k))$ is the initialization *of I.*

As an example, in $\Sigma_{59.2}$, let I^A and I^B be place weights for A and B, respectively, with $I^A(x) = \{f(x)\}$ for each $x \in U$ and $I^B(y) = \{y\}$ for each $y \in V$. In the sequel we show

$$\{I^A, I^B\} \text{ is a trap of } \Sigma_{59.2}. \tag{2}$$

To this end, we first observe that the occurrence modes of both transitions a and b are given by the set U. Then for each $m \in U$:

$$
\begin{aligned}
I^A(m(A,a)) \cup I^B(m(B,a)) &= I^A(\{m\}) \cup I^B(\{f(m)\}) \\
&= \{f(m)\} \cup \{f(m)\} \\
&= \{f(m)\} \\
&= \{f(m)\} \cup \emptyset \\
&= I^A(\{m\}) \cup I^B(\emptyset) \\
&= I^A(m(a,A)) \cup I^B(m(a,B)).
\end{aligned}
$$

Likewise, for the transition b,

$$
\begin{aligned}
I^A(m(A,b)) \cup I^B(m(B,b)) &= I^A(\{m\}) \cup I^B(\{f(m)\}) \\
&= \{f(m)\} \cup \{f(m)\} \\
&= \{f(m)\} \\
&= \emptyset \cup \{f(m)\} \\
&= I^A(m(b,A)) \cup I^B(m(b,B)).
\end{aligned}
$$

Finally, for $t = c$ and $t = d$,

$$I^A(m(A,t)) \cup I^B(m(B,t)) = I^A(\emptyset) \cup I^B(\emptyset)$$
$$= \emptyset \cup \emptyset$$
$$= \emptyset$$
$$\subseteq I^A(m(t,A)) \cup I^B(m(t,B)).$$

Hence, $\{I^A, I^B\}$ is in fact a trap of $\Sigma_{59.2}$. Its initialization is $I^A(U) \cup I^B(\emptyset) = f(U) \cup \emptyset = V$. An initialized trap in fact provides a valid Σ-inequality:

61.2 Theorem. *Let Σ be a system net, let $p_1, \ldots, p_k \in P_\Sigma$, and for $j = 1, \ldots, k$, let I^j be a place weight of Σ. Let $\{I^1, \ldots, I^k\}$ be a trap of Σ with initialization B. Then the inequality*

$$I^1(p_1) \cup \cdots \cup I^k(p_k) \geq B$$

holds in Σ.

Proof. i. Let $r \xrightarrow{t,m} s$ be a step of Σ. Then

$$\bigcup_{j=1}^{k} I^j(s(p_j)) = \bigcup_{j=1}^{k} I^j((r(p_j) \setminus m(p_j, t)) \cup m(t, p_j)) \qquad \text{by Def. 16.3}$$

$$= \bigcup_{j=1}^{k} I^j(r(p_j) \setminus m(p,t)) \cup \bigcup_{j=1}^{k} I^j(m(t,p_j)) \qquad \text{by rules on sets}$$

$$\supseteq \bigcup_{j=1}^{k} (I^j(r(p_j)) \setminus I^j(m(p,t))) \cup \bigcup_{j=1}^{k} I^j(m(t,p_j)) \qquad \text{by rules on sets}$$

$$\supseteq (\bigcup_{j=1}^{k} I^j(r(p_j)) \setminus \bigcup_{j=1}^{k} I^j(m(p,t))) \cup \bigcup_{j=1}^{k} I^j(m(t,p_j))$$

$$\text{by rules on sets}$$

$$= \bigcup_{j=1}^{k} I^j(r(p_j)) \cup \bigcup_{j=1}^{k} I^j(m(t,p_j)) \qquad \text{by Def. 61.1}$$

$$\supseteq \bigcup_{j=1}^{k} I^j(r(p_j)) \qquad \text{by rules on sets.}$$

ii. Now let s be a reachable state of Σ. Then there exists an interleaved run of Σ formed $s_0 \xrightarrow{t_1, m_1} s_1 \xrightarrow{t_2, m_2} \cdots \xrightarrow{t_l, m_l} s_l$ with $s_l = s$. Then $\bigcup_{j=1}^{k} I^j(s_0(p_j)) \supseteq B$, by Def. 61.1. Then for each $i = 1, \ldots, l$, $\bigcup_{j=1}^{k} I^j(s_i(p_j)) \supseteq B$, by i and induction on i. Then the case of $i = l$ implies the proposition. $\square$

Proof of traps can be mimicked symbolically in term-inscribed system nets. To this end, place weights I, and functions $\tilde{\beta}$ assigned to arcs β, are represented symbolically as described in Sect. 60. The function $I \circ \beta$ can then be represented symbolically by the multiset term

$$\tau = I^p[\tilde{\beta}/p] \qquad (3)$$

in analogy to (2) of Sect. 60. Union of functions then can be expressed by set union of singleton sets $\{\tau\}$. Each valuation of the variable p in τ by some $m \in \mathcal{A}_p$ then describes the item $I^p \circ \widetilde{\beta}(m) = I^p(\widetilde{\beta}(m))$. $\mathbb{O}$ denotes the function that returns no value at all.

As an example, the trap in (2) can be verified symbolically as follows:

$$
\begin{aligned}
I^A \circ \widetilde{(A,a)} \cup I^B \circ \widetilde{(B,a)} &= f(A) \circ \widetilde{x} \cup B \circ \widetilde{f(x)} \\
&= f(A)[x/A] \cup B[f(x)/B] \\
&= \{f(x)\} \cup \{f(x)\} \\
&= \{f(x)\} \\
&= \{f(x)\} \cup \mathbb{O} \\
&= f(A)[x/A] \cup \mathbb{O} \\
&= I^A \circ \widetilde{x} \cup I^B \circ \mathbb{O} \\
&= I^A \circ \widetilde{(a,A)} \cup I^B \circ \widetilde{(a,B)}.
\end{aligned}
$$

Likewise,

$$
\begin{aligned}
I^A \circ \widetilde{(A,b)} \cup I^B \circ \widetilde{(B,b)} &= f(A) \circ \widetilde{x} \cup B \circ \widetilde{f(x)} \\
&= \{f(x)\} \\
&= \mathbb{O} \cup B[f(x)/B] \\
&= I^A \circ \mathbb{O} \cup I^B \circ \widetilde{f(x)} \\
&= I^A \circ \widetilde{(b,A)} \cup I^B \circ \widetilde{(b,B)}.
\end{aligned}
$$

Finally, for $t = c$ or $t = d$,

$$
\begin{aligned}
I^A \circ \widetilde{(A,t)} \cup I^B \circ \widetilde{(B,t)} &= \mathbb{O} \cup \mathbb{O} \\
&= \mathbb{O} \\
&\subseteq I^A \circ \widetilde{(t,A)} \cup I^B \circ \widetilde{(t,B)}.
\end{aligned}
$$

The initialization of the trap $\{I^A, I^B\}$ is given symbolically by

$$
\begin{aligned}
I^A(U) \cup I^B(\emptyset) &= f(A)[U/A] \cup B[\emptyset/B] \\
&= f(U) \cup \emptyset \\
&= f(U) \\
&= V.
\end{aligned}
$$

62   State Properties of Variants of the Philosopher System

62.1   State properties of nondeterministic philosophers

We start with state properties of the philosophers system, as considered in Sects. 19 and 20. Figure 62.1 redraws $\Sigma_{19.1}$, with renamed places. It shows

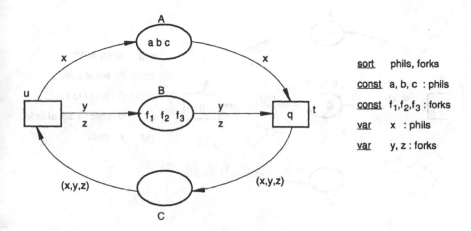

Figure 62.1. System schema for $\Sigma_{19.1}$

the case of a philosopher taking *any* two forks. An obvious place invariant then is

$$A + pr_1(C) = a + b + c, \tag{1}$$

confirming that each philosopher is either thinking or eating. The place invariant

$$B + pr_2(C) + pr_3(C) = f_1 + f_2 + f_3 \tag{2}$$

states that each fork is either available or in use.

The places A and B are quite loosely connected: Each philosopher corresponds to *any* two forks, hence it is just the *number* of philosophers at A and the number of forks at B that can be combined in a place invariant covering A and B. More precisely, philosophers count twice as much as forks do:

$$2|A| - |B| = 3. \tag{3}$$

62.2   State properties in the context of set-valued functions

Figure 62.2 shows a system schema, with each philosopher x taking a fixed set $\Phi(x)$ of forks. $\Sigma_{20.3}$ is an instantiation of this schema. Place invariants of $\Sigma_{62.2}$ are easily gained and interpreted:

$$A + C = P \tag{4}$$

states that each philosopher is either thinking or eating.

$$B + \Phi(C) = G \tag{5}$$

states that each fork is either available or in use by exactly one philosopher, and

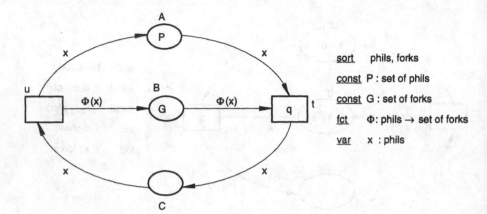

Figure 62.2. System schema for $\Sigma_{20.3}$

$$\Phi(A) - B = \Phi(P) - G \tag{6}$$

states that each philosopher corresponds to the set of his or her forks.

62.3   State properties of the drinking philosophers

Finally, Fig. 62.3 shows a system schema for the drinking philosophers. $\Sigma_{20.5}$ provides an instantiation of this schema.

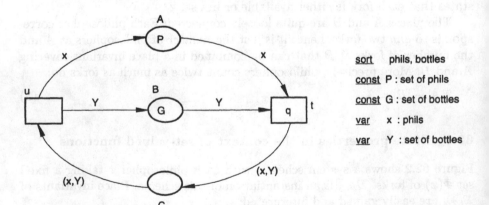

Figure 62.3. System schema for $\Sigma_{20.5}$

Its matrix, initial state, and two place invariants are given in Fig. 62.4. The place invariant I_1 yields the equation

	t	u	M_0	I_1	I_2
A	$-x$	x	P	A	
B	$-Y$	Y	G		B
C	(x,Y)	$-(x,Y)$		$pr_1(C)$	$pr_2(C)$

Figure 62.4. Matrix, initial state, and two place invariants of the drinking philosophers system, $\Sigma_{62.3}$

$$A + pr_1(C) = P, \tag{7}$$

stating that each philosopher is either thinking or eating. Likewise, I_2 yields

$$B + pr_2(C) = G, \tag{8}$$

stating that each bottle is either available or in use. There is no place invariant connecting A and B.

XI. Interleaved Progress of System Nets

Two progress operators have been suggested for elementary system models: the interleaved progress operator $\mapsto$ ("*leads to*") and the concurrent progress operator $\hookrightarrow$ ("*causes*"). They both can be adapted canonically to the case of advanced system nets. The *causes* operator will turn out more important because of its ability for *parallel composition*. In analogy to elementary system nets, we start with progress on interleaved runs.

63 Progress on Interleaved Runs

In analogy to Sect. 44, a progress property $p \mapsto q$ (*p leads to q*) is constructed from two state properties p and q. Now, p and q are first-order state properties, as defined in Sect. 57. Again, as in Sect. 44, $p \mapsto q$ holds in an interleaved run w if each p-state of w is followed by a q-state. $p \mapsto q$ holds in a system net Σ if $p \mapsto q$ holds in each of its interleaved runs. Technically, leads-to formulas are constructed from state formulas:

63.1 Definition. *Let A be a structure, let X be a set of A-sorted variables, let P be a set of symbols, and let $p, q \in \mathcal{F}(A, X, P)$ be state formulas. Then the symbol sequence $p \mapsto q$ (p leads to q) is a first-order leads-to formula.*

Leads-to formulas are interpreted over interleaved runs and over system nets:

63.2 Definition. *Let Σ be a net that is term-inscribed over a structure A and a set X of variables. Let $p, q \in \mathcal{F}(A, X, P_\Sigma)$ and let w be an interleaved run of Σ.*

i. *For an argument u of X let $w \models (p \mapsto q)(u)$ iff for each $p(u)$-state with index i, there exists a $q(u)$-state with index $j \geq i$.*

ii. *$p \mapsto q$ is said to hold in w (written $w \models p \mapsto q$) iff for each argument u of X, $w \models (p \mapsto q)(u)$.*

iii. *$p \mapsto q$ is said to hold in Σ (written $\Sigma \models p \mapsto q$) iff $w \models p \mapsto q$ for each interleaved run w of Σ.*

As an example, in Fig. 63.1 the formula $A.u \wedge A.v \mapsto C.f(u, v)$ is true.

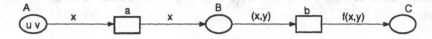

Figure 63.1. $A.u \wedge A.v \mapsto C.f(u,v)$

64   Interleaved Pick-up
and Proof Graphs for System Nets

The pick-up rule for es-nets, as stated in Sect. 45, is canonically extended to system nets. The only slightly nontrivial new notion is the postset $s^\bullet$ of a state s of a system net Σ. In fact, $s^\bullet$ contains *actions* of transitions of Σ. More precisely, an action m of a transition t is in $s^\bullet$ if occurrence of m reduces the token load of some place p, i.e., if $m(p,t) \neq \emptyset$.

64.1 Definition. *Let Σ be a system net and let s be a state of Σ.*

i. *s is* progress prone *iff s enables at least one action of some progressing transition of Σ.*

ii. *Let $t \in T_\Sigma$ and let m be an action of t. s prevents m iff $\Sigma \models \hat{s} \to \neg m(p,t)$.*

iii. *Let $t \in T_\Sigma$ and let m be an action of t. $m \in s^\bullet$ if for some place p of Σ, $s(p) \cap m(p,t) \neq \emptyset$.*

iv. *A set M of actions of some transitions of Σ is a* change set *of s if $M \neq \emptyset$ and s prevents each $m \in s^\bullet \setminus M$.*

The following theorem describes the most general case for picking up leads-to formulas from the static structure of a system net: Each change set of a progress prone state s yields a leads-to formula:

64.2 Theorem. *Let Σ be a system net, let s be a progress prone state, and let M be a change set of s. Then*

$$\Sigma \models s \mapsto \bigvee_{m \in M} \text{eff}(s,m).$$

Proof of this theorem follows the proof of Theorem 45.5 and is left as an exercise for the reader.

64.1   Pick-up patterns

Rules for picking up valid *leads-to* formulas from term-inscribed nets will be presented in the sequel. A most general, fully fledged syntactical pick-up rule, i.e., a syntactical representative of Theorem 64.2, is technically complicated and unwieldy. Some typical patterns will be considered instead, sufficient for verifying an overwhelming majority of case studies.

We start with forward branching places that lead to a disjunction:

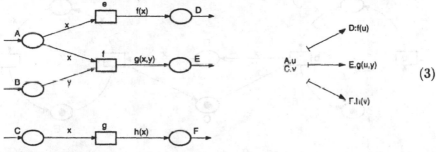

$$A.x \mapsto B.f(x) \vee C.g(x). \qquad (1)$$

Synchronization is as expected:

$$A.x \wedge B.y \mapsto C.f(x,y). \qquad (2)$$

More generally, and with the picked-up formula written as a proof graph,

$$(3)$$

In case additionally $A.u \to \neg B.y$, all actions formed $f(u,w)$ are ruled out and one may pick up

$$(4)$$

Summing up, the interleaved pick-up rule of Sect. 45 canonically generalizes to system nets and will be used accordingly.

64.2 Proof graphs

Proof graphs for interleaved progress of system nets can be constructed in strict accordance with the case of elementary system nets, as introduced in

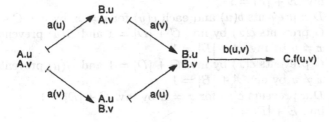

Figure 64.1. Proof graph for $\Sigma_{63.1} \models A.u \wedge A.v \mapsto C.f(u,v)$

Sects. 46 and 47. We refrain from a formal definition here; the general case can easily be derived from the proof graph for $\Sigma_{63.1} \models A.u \wedge A.v \mapsto C.f(u,v)$, given in Fig. 64.1.

65   Case Study: Producer/Consumer Systems

We are now prepared to show for producer/consumer systems that each producer item will eventually be consumed. Figure 65.1 shows a system schema,

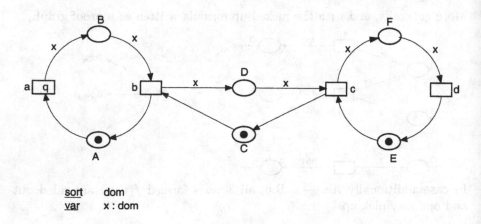

$$\begin{array}{ll} \underline{\text{sort}} & \text{dom} \\ \underline{\text{var}} & \text{x : dom} \end{array}$$

Figure 65.1. System schema for producer/consumer systems

with instantiation as in $\Sigma_{15.5}$. Each item *ready to be delivered* should eventually become *ready to be consumed*. In terms of $\Sigma_{65.1}$ this reads

$$B.u \mapsto F.u. \tag{1}$$

Figure 65.2 shows a proof graph for (1). Its nodes are justified as follows:

node 1: inv. $C + |D| = 1$
node 2: inv. $E + |F| = 1$
node 3: $D.x$ prevents $b(u)$ by inv. $C + |D| = 1$ and $F.x$ prevents $c(x)$ by inv. $E + |F| = 1$
node 4: $D.x$ prevents $b(u)$ and each $c(y)$ for $y \neq x$, by inv. $C + |D| = 1$
node 5: C prevents $c(x)$ by inv. $C + |D| = 1$ and $B(u)$ prevents $b(x)$ for $x \neq u$ by inv. $A + |B| = 1$
node 6: C prevents $c(x)$ by inv. $C + |D| = 1$ and $B(u)$ prevents $b(x)$ for $x \neq u$ by inv. $A + |B| = 1$
node 7: $D.u$ prevents $c(x)$ for $x \neq u$ by inv. $C + |D| = 1$
node 8: inv. $E + |F| = 1$
node 9: $F.x$ prevents $c(u)$ by inv. $E + |F| = 1$.

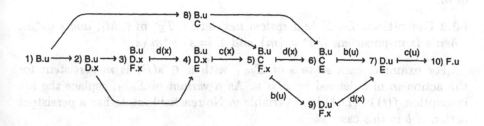

Figure 65.2. Proof graph for $\Sigma_{65.1} \models B.u \mapsto F.u$

66 How to Pick up Fairness

A pick-up rule for leads-to properties is constructed in the sequel that exploits the assumption of fairness of actions. Some technicalities are required first, including the pre- and postsets of actions, and persistence of states. The postset $s^\bullet$ of a state s has already been defined in Sect. 64.1.

66.1 Definition. *Let Σ be a system net, let $t \in T_\Sigma$, and let $m \in M_t$ be an action of t.*

i. *The preset $^\bullet m$ and the postset $m^\bullet$ of m are states of Σ, defined for each place $p \in P_\Sigma$ by $^\bullet m(p) = m(p,t)$ and $m^\bullet(p) = m(t,p)$, respectively.*
ii. *For two states r and s, let $r \setminus s$ be the state defined for each place p of Σ by $(r \setminus s)(p) := r(p) \setminus s(p)$.*

As an example, in Fig. 66.1, $x = u$ defines an action m of b, with $^\bullet m(B) = \{u\}$, $^\bullet m(D) = \{v\}$, and $^\bullet m(A) = {}^\bullet m(C) = {}^\bullet m(E) = \emptyset$. A substate s is

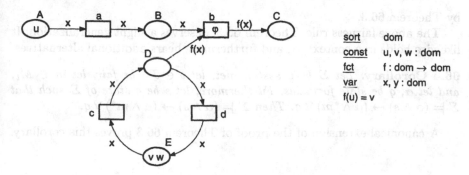

Figure 66.1. s is m-persistent, with $u \in s(B)$ and $m(x) = u$

persistent with respect to an action m if s can be changed only by occurrence of m:

66.2 Definition. *Let Σ be a system net, let $t \in T_\Sigma$, $m \in M_t$, and $s \subseteq {}^\bullet m$. Then s is m-persistent if $s^\bullet = \{m\}$ and $\Sigma \models s \mapsto {}^\bullet m \setminus s$.*

For example, each state s of $\Sigma_{66.1}$ with $u \in s(B)$ is m-persistent for the action m of b defined by $x = u$. As a variant of $\Sigma_{66.1}$, replace the arc inscription $f(t)$ of (D, b) by a variable y. No reachable state has a persistent action of b in this case.

An action m of a fair transition will occur at each m-persistent state:

66.3 Theorem. *Let Σ be a system net, let $t \in T_\Sigma$ be fair, let $m \in M_t$, and let s be a m-persistent state. Then $\Sigma \models s \mapsto m^\bullet$.*

Proof. Let $w = s_0 \xrightarrow{(t_1, m_1)} s_1 \xrightarrow{(t_2, m_2)} s_2 \ldots$ be an interleaved run of Σ. Let s_k be an s-state, i.e., $s_k \models s$. Then $t_{k+1} = (t, m)$ or $s_{k+1} \models s$ because $s^\bullet = \{m\}$. Furthermore, there exists an ${}^\bullet m$-state $s_{l'}$, with $l' \geq k$, because $\Sigma \models s \mapsto {}^\bullet m$. Let l be the smallest such index. Then $s_{l-1} \models {}^\bullet m$. Hence

$$\text{for some } l > k, \ t_l = (t, m) \text{ or } s_l \models s. \tag{1}$$

To show $w \models s \mapsto m^\bullet$, let s_k be an s-state. By iteration of (1), either $t_l = (t, m)$ for some $l > k$ (and hence $s_{l+1} \models s$), or there exists an infinite sequence of s-states. But the latter case is ruled out due to the assumption of fairness for t.   □

Returning to $\Sigma_{66.1}$, the proof graph

$$\text{B.u} \longrightarrow \text{E.v} \longmapsto \text{D.v} \tag{2}$$

proves $B.u \mapsto D.v$, i.e., for each state s with $u \in s(B)$ and each action m of b with $m(x) = u$, $s \models {}^\bullet m \setminus s$. Each such state is m-persistent. Hence

$$\Sigma_{66.1} \models B.u \mapsto C.f(u) \tag{3}$$

by Theorem 66.3.

The above fairness rule, Theorem 66.3, deserves a slight generalization: It likewise holds in a context, α, and furthermore bears additional alternatives.

66.4 Corollary. *Let Σ be a system net, let $t \in T_\Sigma$ be fair, let $m \in M_t$, and let α, q be state formulas. Furthermore, let s be a state of Σ such that $\Sigma \models (\alpha \wedge s) \mapsto (\alpha \wedge {}^\bullet m) \vee q$. Then $\Sigma \models (\alpha \wedge s) \mapsto (\alpha \wedge m^\bullet) \vee q$.*

A canonical extension of the proof of Theorem 66.3 proves this corollary.

XII. Concurrent Progress of System Nets

The above interleaving-based progress operator for advanced system nets is now complemented by a concurrency-based operator $\hookrightarrow$, in analogy to concurrent progress of elementary system nets, as discussed in Chap. IX.

67 Progress of Concurrent Runs

First-order causes formulas are constructed from state formulas as defined in Def. 57.1, and the elementary *causes* operator from Def. 50.2.

67.1 Definition. *Let A be a structure, let X be a set of A-sorted variables, let P be a set of symbols, and let $p, q \in \mathcal{F}(A, X, P)$ be state formulas. Then the symbol sequence $p \hookrightarrow q$ ("p causes q") is a first-order causes formula.*

Causes formulas are interpreted over concurrent runs and over system nets:

67.2 Definition. *Let Σ be a net that is term-inscribed over a structure A and a set X of variables. Let $p, q \in \mathcal{F}(A, X, P_\Sigma)$ and let K be a concurrent run of Σ.*

- *i. For an argument u of X, let $K \models (p \hookrightarrow q)(u)$ iff to each reachable $p(u)$-state C of K there exists a $q(u)$-state D of K that is reachable from C.*
- *ii. $p \hookrightarrow q$ is said to hold in K (written $K \models p \hookrightarrow q$) iff for each argument u of X, $K \models (p \hookrightarrow q)(u)$.*
- *iii. $p \hookrightarrow q$ is said to hold in Σ (written $\Sigma \models p \hookrightarrow q$) iff $K \models p \hookrightarrow q$ for each concurrent run K of Σ.*

As an example, $A.\{u, v\} \hookrightarrow B.\{u, v\}$ holds in

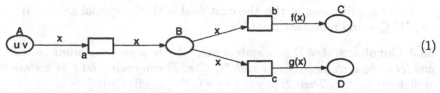

$$(1)$$

As discussed in Lemmas 50.3 and 50.4, properties of the propositional causes operator likewise apply to the first-order causes operator:

67.3 Lemma. *Let Σ be a system net that is term-inscribed over a structure $\mathcal{A}$ and let $p, q \in \mathrm{sf}(\mathcal{A}, X, P_\Sigma)$.*

   i. $\Sigma \models p \hookrightarrow p$.
  ii. If $\Sigma \models p \hookrightarrow q$ and $\Sigma \models q \hookrightarrow r$ then $\Sigma \models p \hookrightarrow r$.
 iii. If $\Sigma \models p \hookrightarrow r$ and $\Sigma \models q \hookrightarrow r$ then $\Sigma \models (p \vee q) \hookrightarrow r$.
  iv. If $\Sigma \models p \mapsto q$ then $\Sigma \models p \hookrightarrow q$.
   v. If q includes no logical operator and $\Sigma \models p \hookrightarrow q$ then $\Sigma \models p \mapsto q$.

68   The Concurrent Pick-up Rule

A rule to pick up *causes* properties from a system net is now derived, in an entirely *semantical* framework. The problem of picking up causes formulas from a term-inscribed representation of system nets, is postponed to the next section.

We start with some properties and notations of states of system nets.

68.1 Definition. *Let Σ be a system net and let r, s be two states of Σ.*

   i. The state $r \cup s$ of Σ is defined for each place $p \in P_\Sigma$ by $(r \cup s)(p) := r(p) \cup s(p)$.
  ii. Let $r \subseteq s$ iff for each place $p \in P_\Sigma$, $r(p) \subseteq s(p)$.
 iii. r is disjoint with s iff for each place $p \in P_\Sigma$, $r(p) \cap s(p) = \emptyset$.
  iv. For an action m of some transition t, let $^\bullet m$ be a state of Σ, defined for each place $p \in P_\Sigma$ by $^\bullet m(p) = m(p, t)$. For a set M of actions, let $^\bullet M$ be the state defined for each $p \in P_\Sigma$ by $^\bullet M(p) = \bigcup \{ m(p) \mid m \in M \}$.

Change sets of system nets, as defined in Def. 64.1 for interleaved progress, can likewise be used for concurrent progress properties:

68.2 Theorem. *Let Σ be a system net and let r, s be states of Σ. Assume s is progress prone, and let $U = V \cup W$ be a change set of s, with $^\bullet V \subseteq s$ and r disjoint with $^\bullet V$. Then $\Sigma \models r \cup s \hookrightarrow (r \cup \bigvee_{u \in V} \mathrm{eff}(s, u)) \vee (\bigvee_{u \in W} \mathrm{eff}(r \cup s, u))$.*

Proof of this theorem follows proof of Theorem 51.1 and is left as an exercise for the reader.

Many applications of this theorem deal with the special case of $W = \emptyset$, i.e., $^\bullet U \subseteq s$ and r disjoint from $^\bullet U$:

68.3 Corollary. *Let Σ be a system net, let s be a progress prone state of Σ, and let U be a change set of s with $^\bullet U \subseteq s$. Furthermore, let r be a state that is disjoint with s. Then $\Sigma \models r \cup s \hookrightarrow r \cup (\bigvee_{u \in U} \mathrm{eff}(s, u))$.*

69   Pick-up Patterns and Proof Graphs

In analogy to the pattern of Sect. 64.1, valid *causes* formulas can be picked up from term-inscribed nets with the help of pick-up patterns, as suggested in the sequel.

We stick to *elementary* formulas, avoiding the negation operator ¬.

69.1 Notations. *Let* Σ *be an es-net that is term-inscribed over a structure* $\mathcal{A}$ *and a set* X *of variables.*

  i. *A state formula* p *in* $\mathrm{sf}(\mathcal{A}, X, P_\Sigma)$ *is elementary if the negation symbol* ¬ *does not occur in* p.
  ii. *For a place* $p \in P_\Sigma$ *and a state formula* q, *we write* $p \not\subseteq q$ *if* p *does not occur in* q.

In case $p \not\subseteq q$, the place p, considered as a state, is disjoint to state q.

69.1   The elementary pattern

Most elementary is the case of a forward unbranched place, A, linked to a backward unbranched transition, a:

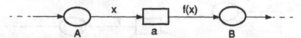

Let α be an elementary state formula with $A, B \not\subseteq \alpha$.

  i. $\alpha \wedge A.x \overset{a(x)}{\hookrightarrow} \alpha \wedge \neg A.x \wedge B.f(x)$
  ii. $\alpha \wedge A.(x + U) \overset{a(x)}{\hookrightarrow} \alpha \wedge A.U \wedge B.f(x)$
  iii. $\alpha \wedge A = U \overset{a(U)}{\hookrightarrow} \alpha \wedge A = 0 \wedge B.f(U)$

69.2   The alternative pattern

The typical free choice alternative is likewise easy:

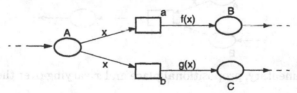

Let α be an elementary state formula with $A, B, C \not\subseteq \alpha$.

  i. $\alpha \wedge A.x \hookrightarrow \alpha \wedge \neg A.x \wedge (B.f(x) \vee C.g(x))$
  ii. $\alpha \wedge A.(x + U) \hookrightarrow \alpha \wedge A.U \wedge (B.f(x) \vee C.g(x))$
  iii. $\alpha \wedge A = U \hookrightarrow \alpha \wedge A = 0 \wedge B.f(V) \wedge C.g(W) \wedge U = V \cup W$

69.3   The synchronizing pattern

Synchronization of places without alternatives goes as can be expected:

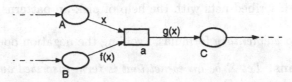

Let α be an elementary state formula with $A, B, C \notin \alpha$.

i. $\alpha \wedge A.x \wedge B.f(x) \overset{a(x)}{\hookrightarrow} \neg A.x \wedge \neg B.f(x) \wedge C.g(x)$

ii. $\alpha \wedge A = U \wedge f(A) \subseteq B \overset{a(A)}{\hookrightarrow} \alpha \wedge A = \emptyset \wedge C.g(U)$

69.4   The pattern for alternative synchronization

Choice between synchronized transitions yields important patterns:

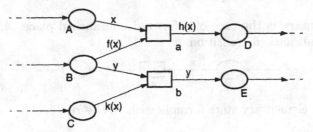

Let α be an elementary state formula with $A, \ldots, E \notin \alpha$.

i. $\alpha \wedge A.x \wedge B.f(x) \hookrightarrow \alpha \wedge D.h(x) \vee E.f(x)$

ii. Let $A.x$ prevent $b(f(x))$. Then $\alpha \wedge A.x \wedge B.f(x) \overset{a(x)}{\hookrightarrow} D.h(x)$

iii. $A = U \wedge f(A) \subseteq B \hookrightarrow D.h(V) \wedge E.f(W) \wedge U = V \cup W$

A frequent special case of this pattern is

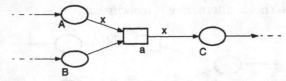

with B an elementary, propositional place and x varying over the set U. Then

i. $\alpha \wedge A.x \wedge B \hookrightarrow \alpha \wedge \bigvee_{y \in U} C.y$

ii. Let inv $|A| \leq 1$ be a valid inequality. Then $A.x \wedge B \hookrightarrow C.x$.

69.5   The pattern for synchronized alternatives

There frequently occur two or more alternatives that are synchronized along
a backwards branched place:

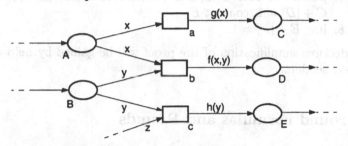

Let α be an elementary state formula with $A, \ldots, E \notin \alpha$.

i. $\alpha \wedge A.x \wedge B.y \hookrightarrow \alpha \wedge D.f(x,y) \vee C.g(x) \vee E.h(y)$
ii. Let $A.x$ prevent $c(y,z)$ and let $B.y$ prevent $a(x)$. Then $A.x \wedge B.y \hookrightarrow D.f(x,y)$.

69.6   Proof graphs for causes formulas

Based on Lemma 67.3, proof graphs for causes formulas can be constructed
as usual.

As an example we turn back to the producer/consumer system. Figure 65.2 shows that each produced item will eventually be consumed; technically, $B.u \hookrightarrow F.u$ for each item u. As an alternative we observe with
Lemma 67.3(v) that it was sufficient to prove $B.u \hookrightarrow F.u$ instead. Figure 69.1

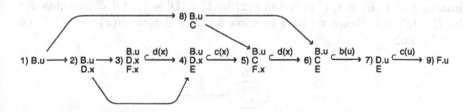

Figure 69.1. $\Sigma_{65.1} \models B.u \hookrightarrow F.u$

shows a corresponding proof graph. In comparison to Fig. 65.2, one node has
vanished. More important is the simplification in the nodes' justification:

node 1: inv. $C + |D| = 1$,
node 2: inv. $E + |F| = 1$,
node 3: pattern of Sect. 69.1, context $B.u \wedge D.x$,
node 4: pattern of Sect. 69.4, $D.x$ prevents $c(y)$ for each $y \neq x$ by inv.
$C + |D| = 1$, context $B.u$,

node 5: pattern of Sect. 69.1, context $B.u \wedge C$,

node 6: pattern of Sect. 69.4, $B.u$ prevents each $b(x)$ for $x \neq u$ by inv.
$A + |B| = 1$, context E,

node 7: pattern of Sect. 69.4, $D.u$ prevents each $c(x)$ for $x \neq u$ by inv.
$C + |D| = 1$, context E,

node 8: inv. $E + |F| = 1$.

Further decisive simplification of the proof will be gained by help of *rounds*
in the next section.

70   Ground Formulas and Rounds

Ground formulas and rounds of elementary system nets are now canonically
extended to advanced system nets:

70.1 Definition. *Let Σ be a system net and let p be a state formula of Σ.
Then p is a* ground formula *of Σ if $\Sigma \models \text{true} \hookrightarrow p$.*

70.2 Theorem. *Let Σ be a system net and let s be a state of Σ. Then s is
a ground formula of Σ iff $\Sigma \models a_\Sigma \hookrightarrow s$ and there exists a change set U of s
such that for each $u \in U$, $\Sigma \models \text{eff}(s, u) \hookrightarrow s$.*

As an example, for the producer/consumer system in Fig. 65.1 we prove

$$ACE \text{ is a ground formula of } \Sigma_{65.1}. \tag{1}$$

The first condition of Theorem 70.2, $\Sigma \models a_\Sigma \hookrightarrow ACE$, is trivially fulfilled,
as $a_\Sigma = ACE$. For the second condition we observe that $\{a(u) \mid u \in \text{dom}\}$ is
a change set of ACE, because for all $x \in \text{dom}$, A prevents $b(x)$ by the place
invariant $A + |B| = 1$, C prevents $c(x)$ by $C + |D| = 1$, and E prevents $d(x)$
by $E + |F| = 1$. Hence we have to show for all $x \in \text{dom}$: $B(x) \hookrightarrow ACE$. The
proof graph

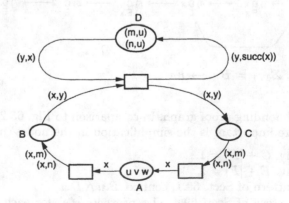

Figure 70.1. Renamed distributed request service $\Sigma_{19.3}$

$$1) B.x \wedge C \wedge E \overset{b(x)}{\hookrightarrow} 2) A \wedge D.x \wedge E \overset{c(x)}{\hookrightarrow} 3) A \wedge C \wedge F.x \overset{d(x)}{\hookrightarrow} 4) A \wedge C \wedge E \qquad (2)$$

shows this property. Its nodes are justified as follows, with all formulas due to the elementary pick-up pattern of Sect. 69.1:

> node 1: pattern of Sect. 69.4, $B.x$ prevents $b(y)$ for $y \neq x$ by inv. $A + |B| = 1$, context E,
> node 2: pattern of Sect. 69.4, $D.x$ prevents $c(y)$ for $y \neq x$ by inv. $C + |D| = 1$, context A,
> node 3: pattern of Sect. 69.4, context $A \wedge C$.

As a further example we show that the initial state of the Distributed Request Service of Fig. 19.3 is a ground state. Figure 70.1 renames this system. As a technical simplification, pairs (x, y) will be written xy; hence we have to show

$$D.mu, nu \wedge A.u, v, w \text{ is a ground formula of } \Sigma_{70.1}. \qquad (3)$$

The set $\{a(x) \mid x \in \{u, v, w\}\}$ apparently is a change set of $a_{\Sigma_{70.1}}$. Furthermore, let $x, y, z \in \{u, v, w\}$ be pairwise different. Then (3) follows with Theorem 70.2 from the following proof graph:

$$
\begin{aligned}
\text{eff}(a_\Sigma, a(x)) = \ &1) D.mu, nu \wedge A.y, z \wedge B.xm, xn \\
&\hookrightarrow 2) D.mu, nu \wedge B.xm, xn, ym, yn, zm, zn \\
&= 3) D.mu, nu \wedge B.um, un, vm, vn, wm, wn \\
&\hookrightarrow 4) D.mv, nv \wedge B.vm, vn, wm, wn \wedge C.mu, nu \\
&\hookrightarrow 5) D.mw, nw \wedge B.wm, wn \wedge C.mu, nu, mv, nv \\
&\hookrightarrow 6) D.mu, nu \wedge C.mu, nu, mv, nv, mw, nw \\
&\hookrightarrow 7) D.mu, nu \wedge A.u, v, w.
\end{aligned}
$$

All nodes are justified by the elementary pick-up pattern of Sect. 69.1.

70.3 Theorem. *Let Σ be a system and let p be a ground formula of Σ. Let s be a state of Σ with $\Sigma \models s \to \neg p$, and let U be a change set of s. Then $\Sigma \models s \hookrightarrow \bigvee_{u \in U} \text{eff}(s, u)$.*

This theorem simplifies proof of *leads-to* formulas in many cases. As an example, Fig. 69.1 provides a proof of $\Sigma_{65.1} \models B.u \hookrightarrow F.u$. This property also follows from the proof graph

$$1) B.u \overset{b(u)}{\hookrightarrow} 2) D.u \overset{c(u)}{\hookrightarrow} 3) F.u. \qquad (4)$$

Its nodes are justified as follows:

> node 1: $B.u$ prevents $b(y)$ for $y \neq x$, and $B.u \to \neg ACE$, by inv. $A + |B| = 1$. Hence the proposition with (1) and Theorem 70.3.
> node 2: $D.u$ prevents $c(y)$ for $y \neq x$, and $D.u \to \neg ACE$, by inv. $C + |D| = 1$. Hence the proposition with (1) and Theorem 70.3.

XIII. Formal Analysis of Case Studies

The case studies of Part B, as introduced in Chaps. IV, V, and VI, are now reconsidered and formally verified.

71 The Asynchronous Stack

71.1 Properties of modules

The central state property of the asynchronous stack $\Sigma_{22.6}$ states that each module M_i is always quiet, or storing two values, or storing no value. In the stack's representation of Fig. 71.1, the equation

$$A_1 + B_1 + C = (1) + \cdots + (n), \tag{1}$$

states this property. Brackets indicate that numbers $1, \ldots, n$ are to be considered as data values, and addition as multiset addition of singleton sets. Proof of (1) is easy: (1) is the equation of the place invariant given in Fig. 71.2.

71.2 Balanced states

A state of the asynchronous stack is *balanced* if each module A_i is at its *quiet* state, storing exactly one value. In terms of Fig. 71.1, a state is *balanced* in case there exist values $u_1, \ldots, u_n$ with

$$A.(1, u_1), \ldots, (n, u_n). \tag{2}$$

A balanced state enables the actions $b_0(u_1)$ and $a_0(v, u_1)$, for all values v. $b_0(u_1)$ pops u_1 out of the stack, yielding the intermediate state

$$C.1 \wedge A.(2, u_2), \ldots, (n, u_n). \tag{3}$$

This state is eventually followed by the balanced state

$$A.(1, u_1), \ldots, (n-1, u_n), (n, \perp), \tag{4}$$

as shown by the following proof graph:

$$C.1 \wedge A.(2, u_2), \ldots, (n, u_n) \overset{b(1, u_2)}{\rightsquigarrow}$$
$$C.2 \wedge A.(1, u_2), (3, u_3), \ldots, (n, u_n) \overset{b(2, u_3)}{\rightsquigarrow}$$

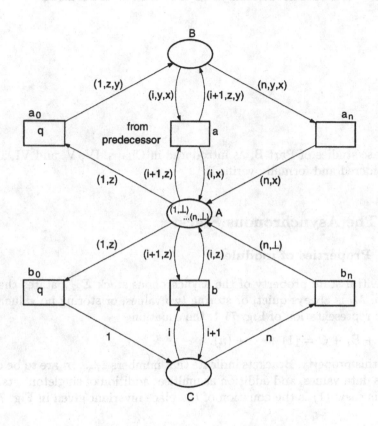

$$
\begin{array}{lll}
\underline{\text{sort}} & \text{value} & \underline{\text{var}} \quad \text{x, y, z : nat} \\
\underline{\text{const}} & \perp : \text{value} & \underline{\text{var}} \quad \text{i : nat} \\
\underline{\text{const}} & \text{n : nat} &
\end{array}
$$

Figure 71.1. Renamed asynchronous stack $\Sigma_{22.6}$

	a_0	a	a_n	b_0	b	b_n	i
A	$-(1,z)$	(i,x) $-(i+1,z)$	(n,x)	$-(1,z)$	(i,z) $-(i+1,z)$	$-(n,\perp)$	$pr_1(\mathbf{A})$
B	$(1,z,y)$	$(i+1,z,y)$ $-(i,y,x)$	$-(n,y,x)$				$pr_1(\mathbf{B})$
C					1	$(i+1)$ $-(i)$	$-n$	$\mathbf{C}$

Figure 71.2. Matrix and place invariant to $\Sigma_{71.1}$

$\vdots$

$$C.i \wedge A.(1, u_2), \ldots, (i-1, u_i), (i+1, u_{i+1}), \ldots, (n, u_n) \overset{b(i, u_{i+1})}{\hookrightarrow}$$

$\vdots$

$$C.n \wedge A.(1, u_2), \ldots, (n-1, u_n) \overset{b_n}{\hookrightarrow}$$
$$A.(1, u_2), \ldots, (n-1, u_n), (n, \bot)$$

The pattern of Sect. 69.4 and the above equation (1) justify this proof graph. Likewise, an action $a_0(v, u_1)$ pushes v into the stack, yielding the intermediate state

$$B.(1, u_1, v) \wedge A.(2, u_2), \ldots, (n, u_n). \tag{5}$$

This state is eventually followed by the balanced state

$$A.(1, v), (2, u_1), \ldots, (n, u_{n-1}), \tag{6}$$

as shown by the following proof graph:

$$B.(1, u_1, v) \wedge A.(2, u_2), \ldots, (n, u_n) \overset{a(1, u_1, v)}{\hookrightarrow}$$
$$D.(2, u_2, u_1) \wedge A.(1, v), (3, u_3), \ldots, (n, u_n) \overset{a(2, u_2, u_1)}{\hookrightarrow}$$

$\vdots$

$$B.(i, u_i, u_{i-1}) \wedge A.(1, v), (2, u_1), \ldots, (i-1, u_{i-2}),$$
$$(i+1, u_{i+1}), \ldots, (n, u_n) \overset{a(i, u_i, u_{i-1})}{\hookrightarrow}$$

$\vdots$

$$B.(n, u_n, u_{n-1}) \wedge A.(1, v), (2, u_1), \ldots, (n-1, u_{n-2}) \overset{a(n, u_{n-1}, u_n)}{\hookrightarrow}$$
$$A.(1, v), (2, u_1), \ldots, (n, u_{n-1}).$$

The pattern of Sect. 69.4 and the above place invariant (1) justify this proof graph.

71.3   A ground formula

The balanced states of $\Sigma_{71.1}$ are characterized by the formula

$$pr_1(A).1, \ldots, n. \tag{7}$$

Given a balanced state s with $A.u_1$, the actions $b_0(u_1)$ and all actions $a_0(v, u_1)$ (for all values v) form a progress set of s. With the above proof graphs and Def. 70.1 it follows that (7) is a ground formula of $\Sigma_{71.1}$. Hence, each reachable state of the asynchronous stack is eventually followed by a balanced state. Furthermore, with the proof graphs above, a push followed by a pop returns the original stack up to the stack's last element, which will contain the undefined element, $\bot$:

$$A.(1, u_1), \ldots, (n, u_n) \overset{a_0(v, u_1)}{\hookrightarrow}$$

$$\vdots$$

$$A.(1,v),(2,u_1),\ldots,(n,u_{n-1}) \overset{b_0(v)}{\hookrightarrow}$$

$$\vdots$$

$$A.(1,u_1),\ldots,(n-1,u_{n-1}),(n,\perp).$$

72 Exclusive Writing and Concurrent Reading

Two algorithms of Sect. 24 are now proven correct. Three properties are to be shown for each of them: exclusive writing, concurrent reading, and evolution. To improve the technical treatment, the two algorithms' places have been relabeled in Figs. 72.1 and 72.3, respectively.

72.1 Proof of exclusive writing and concurrent reading of $\Sigma_{24.2}$

Exclusive writing of $\Sigma_{24.2}$, as redrawn in Fig. 72.1, can easily be shown by help of the place invariant

$$\widetilde{R}(D) + F + K = R \tag{1}$$

with $\widetilde{R}(x) := R$ for each $x \in W$. This invariant immediately implies $|D| \le 1$, i.e., no two writer processes are writing coincidently. It furthermore implies $D = x \rightarrow K = 0$, i.e., if one process is writing, no process is reading. Concurrent reading can easily be demonstrated by means of a

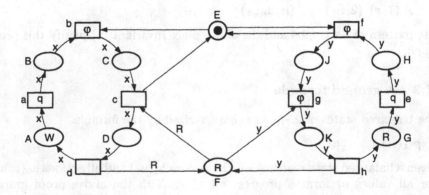

Figure 72.1. Renamed writers/readers $\Sigma_{24.2}$

prefix w of an interleaved run of $\Sigma_{72.1}$. The sequence of actions of w is $e(r_1),\ldots,e(r_n), f(r_1),\ldots,f(r_n), g(r_1),\ldots,g(r_n)$ where $R = \{r_1,\ldots,r_n\}$. w terminates in a state a with $a \models K(R)$, i.e., all reader processes are reading.

72.2 Proof of evolution of writers for $\Sigma_{24.2}$

Evolution of writers of $\Sigma_{24.2}$, as redrawn in Fig. 72.1, reads

$$\Sigma_{72.1} \models B.x \mapsto D.x . \tag{2}$$

Its proof is essentially based on the property

$$\Sigma_{72.1} \models C.x \hookrightarrow D.x \wedge E . \tag{3}$$

This property in turn holds due to the proof graph of Fig. 72.2. Its nodes are

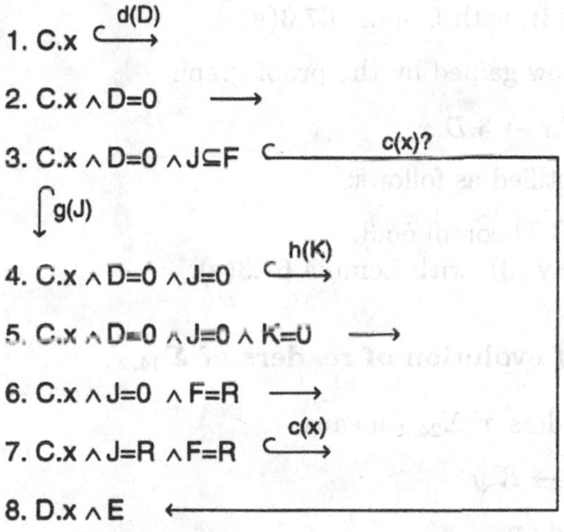

1. $C.x \xrightarrow{d(D)}$

2. $C.x \wedge D{=}0 \longrightarrow$

3. $C.x \wedge D{=}0 \wedge J{\subseteq}F \xhookleftarrow{\quad c(x)? \quad}$

$\left\lfloor g(J) \right.$

4. $C.x \wedge D{=}0 \wedge J{=}0 \xhookrightarrow{h(K)}$

5. $C.x \wedge D{=}0 \wedge J{=}0 \wedge K{=}U \longrightarrow$

6. $C.x \wedge J{=}0 \wedge F{=}R \longrightarrow$

7. $C.x \wedge J{=}R \wedge F{=}R \xhookrightarrow{c(x)}$

8. $D.x \wedge E \longleftarrow$

Figure 72.2. Proof graph for $\Sigma_{72.1}$

justified as follows:

1: pattern of Sect. 69.1 and context $C.x$.
2: inv $\widetilde{R}(D) + F - J - H - G = 0$.
3: $C.x \wedge x \neq z \rightarrow \neg C.z$ with inv $|C| + |E| = 1$, hence $C.x$ prevents $C.z$ for $x \neq z$, hence $C.x \wedge J.y \wedge F.y \hookrightarrow K.y \vee (D.x \wedge E)$ with pattern of Sect. 69.4(i) and action $g(y)$ or $c(x)$, hence $C.x \wedge J \subseteq F \hookrightarrow J.0 \vee (D.x \wedge E)$ with pattern of Sect. 69.4(iii), hence the proposition with context $D = 0$.
4: $K \xhookrightarrow{h(K)} K.0$ by pattern of Sect. 69.1, hence the proposition with context $C.x \wedge J = 0 \wedge D = 0$.
5: inv $\widetilde{R}(D) + F + K = R$.
6: inv $G + H + J + K = R$ and inv $J + \bar{J} = R$, with $\bar{J}$ the complement of J.
7: $C.x \wedge x \neq z \rightarrow \neg C.z$ with inv $|C| + |E| = 1$, hence $C.x$ prevents $c(z)$ for $z \neq x$. $C.x \rightarrow \neg E$ by inv $|C| + |E| = 1$, hence $C.x$ prevents $f(y)$. $\bar{J}.R \rightarrow J.0$ by inv $J + \bar{J} = R$, hence $\bar{J}$ prevents $g(y)$. Hence the proposition with pattern of Sect. 69.4(ii).

Furthermore, proof of (2) requires

$$\Sigma_{72.1} \models B.x \mapsto E. \tag{4}$$

This can be shown by the proof graph

1. B.x $\longrightarrow$ 2. C.z $\longmapsto$ 3. E

Its nodes are justified as follows:

1: inv $|C| + |E| = 1$.
2: property (3), with Lemma 67.3(v).

Proof of (2) is now gained by the proof graph

$$1.B.x \mapsto 2.C.x \mapsto 3.D.x.$$

Its nodes are justified as follows:

1: by (4) and Theorem 66.3.
2: by property (3), with Lemma 67.3(v).

72.3   Proof of evolution of readers of $\Sigma_{24.2}$

Evolution of readers of $\Sigma_{24.2}$ means

$$\Sigma_{72.1} \models H.y \mapsto K.y. \tag{5}$$

Its proof is based on

$$\Sigma_{72.1} \models H.y \mapsto E, \tag{6}$$

to be shown by analogy to (4). Furthermore, we require

$$\Sigma_{72.1} \models J.y \mapsto F.y, \tag{7}$$

which holds due to the proof graph

$$1.J.y \stackrel{d(D)}{\hookrightarrow} 2.D = 0 \stackrel{h(K)}{\hookrightarrow} 3.D = 0 \wedge K = 0 \rightarrow 4.F = R \rightarrow 5.F.y.$$

Its nodes can be justified by analogy to the nodes of the proof graph of (3), left as an exercise to the reader.

Proof of (5) now follows with the proof graph

$$1.H.y \mapsto 2.J.y \mapsto 3.K.y.$$

Its nodes are justified as follows:

1: Theorem 66.3, with (6).
2: Theorem 66.3, with (7).

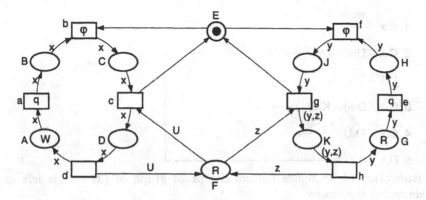

Figure 72.3. Renamed writers/readers $\Sigma_{24.3}$

72.4 Proof of exclusive writing and concurrent reading of $\Sigma_{24.3}$

By analogy to the case of $\Sigma_{24.2}$ exclusive writing of $\Sigma_{24.3}$ as redrawn in Fig. 72.3 can easily be shown by means of the place invariant

$$\widetilde{U}(D) + F + pr_2(K) = U \tag{8}$$

with $\widetilde{U}(x) := U$ for each $x \in W$. This invariant immediately implies $|D| \leq 1$, i.e., no two writer processes are writing coincidently. Furthermore, $D = x \rightarrow K = 0$, i.e., if one process is writing, then no process is reading. Concurrent reading is in fact possible for up to $|U|$ reader processes, in case $|U| \leq |R|$. This can be demonstrated by means of a prefix w of an interleaved run of $\Sigma_{72.3}$. The sequence of actions of w is

$$e(r_1), \ldots, e(r_n), f(r_1), g(r_1, u_1), \ldots, f(r_m), g(r_m, u_m),$$

where $R = \{r_1, \ldots, r_n\}$ and $U = \{u_1, \ldots, u_m\}$. w terminates in a state a with $a \models K(\{(r_1, u_1), \ldots, (r_m, u_m)\})$, i.e., $m = |U|$ reader processes reading.

72.5 Proof of evolution of writing for $\Sigma_{24.3}$

Evolution of writers of $\Sigma_{24.3}$ means

$$\Sigma_{72.3} \models B.x \mapsto D.x. \tag{9}$$

Its proof is essentially based on

$$\Sigma_{72.3} \models C.x \hookrightarrow D.x \wedge E. \tag{10}$$

This property holds due to the proof graph

1. $C.x \xrightarrow{d(D)}$

2. $C.x \wedge D=0 \quad C \xleftarrow{\qquad c(x)? \qquad}$

$\downarrow h(K)$

3. $C.x \wedge D=0 \wedge K=0 \longrightarrow$

4. $C.x \wedge F=U \xrightarrow{c(x)}$

5. $D.x \wedge E \longleftarrow$

Justification of its nodes follows the proof graph of (3) and is left as an exercise to the reader.

Proof of (9) furthermore requires

$$\Sigma_{72.3} \models B.x \mapsto E. \tag{11}$$

This property follows from the proof graph

1. $B.x \longrightarrow$ 2. $C.z \longmapsto$ 3. E

Its nodes are justified as follows:

1: inv $|C| + |E| = 1$.
2: property (10).

Now, (9) follows from the proof graph

$1.B.x \mapsto 2.C.x \mapsto 3.D.x.$

Its nodes are justified as follows:

1: by the fairness rule (Theorem 66.3) with (11).
2: by property (10).

72.6 Proof of evolution of readers of $\Sigma_{24.3}$

Evolution of readers of $\Sigma_{24.3}$ means

$$\Sigma_{72.3} \models H.y \mapsto K.(y,z). \tag{12}$$

Its proof is based on

$$\Sigma_{72.3} \models H.y \mapsto E, \tag{13}$$

to be shown by analogy to (11). Furthermore, we require

$$\Sigma_{72.1} \models J.y \mapsto K.(y,z),$$

which holds due to the proof graph

$$1.J.y \overset{d(D)}{\hookrightarrow} 2.J.y \wedge D = 0 \overset{h(K)}{\hookrightarrow} 3.J.y \wedge D = 0 \wedge K = 0 \tag{14}$$
$$\rightarrow 4.J.y \wedge U = F \hookrightarrow 5.K.(y,z).$$

Its nodes are justified as follows:

1: $D \overset{d(D)}{\hookrightarrow} D.0$ with pattern of Sect. 69.1(iii) and context $J.y$.

2: $K \overset{h(K)}{\hookrightarrow} K.0$ with pattern of Sect. 69.1(iii) and context $J.y \wedge D = 0$.

3: inv $\tilde{U}(D) + F + pr_2(K) = U$.

4: $J.y \wedge y \neq z \rightarrow \neg J.z$ by inv $|E| + |J| + |C| = 1$, hence $J.y$ prevents $g(z)$ for $z \neq y$. $J.y \rightarrow \neg C.x$ by inv $|E| + |J| + |C| = 1$, hence $J.y$ prevents $c(x)$. Hence the proposition with pattern of Sect. 69.4(iii).

This completes the proof of properties of the writer/reader system of Sect. 24.

73 Distributed Rearrangement

Figure 73.1 rewrites $\Sigma_{25.5}$, with renamed places. We first construct a ground formula, *ground*. This formula enables at least one transition, unless the two sets are rearranged. A descending function will show that *ground* will be reachable only finitely often.

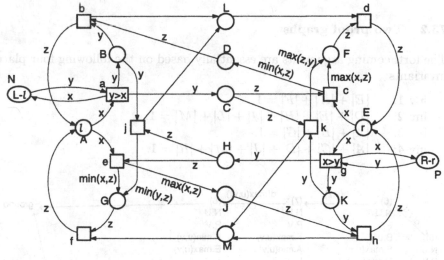

```
const  l,r         : nat
const  L,R         : set of nat
var    x,y,z       : nat
fct    min,max     : nat × nat → nat
(L ∪ {l}) ∩ (R ∪ {r}) = ∅
∃ m ∈ L : max(l,m) = m
```

Figure 73.1. Renamed distributed message passing rearrangement $\Sigma_{25.5}$

73.1   Some basics

We assume a fixed interpretation of the symbols in $\Sigma_{25.5}$, with

$$\text{dom} := L \cup R \tag{1}$$

the set of numbers involved, as initially given. The formula

$$ground := (N + P + A + E = \text{dom}) \wedge A.x \wedge E.y \wedge A < E \tag{2}$$

will turn out to be a ground formula: Both sites carry a test number on A and E, respectively, with $A < E$. All other numbers are collected at N and P, respectively. The degree of disorder at a *ground* state between the two sites is measured by

$$dis := |\{(u,v) \in (N \cup A) \times (P \cup E) \mid u > v\}|, \tag{3}$$

hence the two sites are rearranged in a *ground* state if disorder has disappeared:

$$\text{rearr} := (dis = 0). \tag{4}$$

73.2   Two proof graphs

The forthcoming arguments are essentially based on the following four place invariants:

inv 1: $|E| + |K| + |F| = 1.$
inv 2: $|D| + |E| + |H| + |J| + |L| + |M| = 1.$
inv 3: $|A| + |B| + |G| = 1.$
inv 4: $|A| + |C| + |D| + |J| + |M| + |R| = 1.$

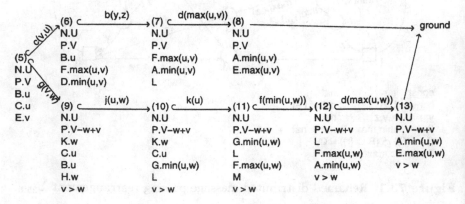

Figure 73.2. Proof graph for $\Sigma_{73.1}$

Figure 73.2 shows a proof graph (its node numbering continues the above numbered lines), with nodes justified as follows (we refrain from explicitly mentioning the respective patterns of Sect. 69):

node 5: $g(v, w)$ is actually a schema for all $w \in V$ with $v > w$. Furthermore, context $N.U \wedge B.u$; $E.v$ excludes $k(u, w)$ for each $w \in$ dom, by inv 1.

node 6: context $N.U \wedge P.V \wedge F.\max(u, v)$; $D.\min(u, v)$ excludes $j(u, w)$ for each $w \in$ dom, by inv 2.

node 7: context $N.U \wedge P.V \wedge A.\min(u, v)$.

node 8: propositional reasoning.

node 9: context $N.U \wedge P.V - w + v \wedge K.w \wedge C.u$; $H.w$ excludes $b(u, z)$ by inv 2; $B.u$ excludes $e(u, w)$ for each $u \in$ dom, by inv 3.

node 10: context $N.U \wedge P.V - w + v \wedge G.\min(u, w) \wedge L$; $K.w$ excludes $c(u, w)$ for each $w \in$ dom by inv 1; $C(u)$ excludes $h(w, u)$ for each $u \in$ dom, by inv 4.

node 11: context $N.U \wedge P.V - w + v \wedge L \wedge F.\max(u, w)$.

node 12: context $N.U \wedge P.V - w + v \wedge L \wedge A.\min(u, w)$.

node 13: propositional reasoning.

Figure 73.3 outlines a proof graph that, symmetrically to Fig. 73.2, swaps the left and the right site of Fig. 73.2.

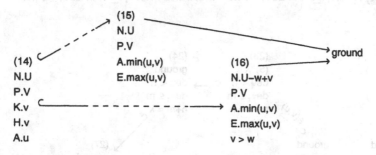

Figure 73.3. Proof graph for $\Sigma_{73.1}$

First we proof that *ground* is in fact a ground formula:

73.3   A ground formula

ground is a ground formula of $\Sigma_{73.1}$.

Proof. i. $s_{\Sigma_{73.3}} \hookrightarrow (5)$ according to the specification in Fig. 73.3. Furthermore, $(5) \hookrightarrow ground$ by Fig. 73.2. Hence $s_{\Sigma_{73.3}} \hookrightarrow ground$.

ii. *ground* prevents each action of e by inv 2 and each action of f by inv 4, hence a progress set of *ground* is given by all actions $a(u, w)$ with $A.u$ and $N.w$ and $w > u$, together with all actions $g(v, w)$ with $E.v$ and $P.w$

and $v > w$. Actions $a(u, w)$ and $g(v, w)$ lead to states shaped as (5) and (14), respectively. The proposition then follows from the proof graphs in Figs. 73.2 and 73.3, and Theorem 70.2.   □

73.4   Proof of rearrangement

In case of ground states, disorder (c.f. (3)) can be characterized in terms of the derivation of the test elements in A and E from the minimum of N and the maximum of P, respectively. To this end, let

$$dev_l := \{w \in N \mid u \in A \wedge w > u\}$$
$$dev_r := \{w \in P \mid u \in E \wedge u < w\}. \tag{17}$$

Then, at each *ground* state holds obviously

$$dis = dev_l + dev_r. \tag{18}$$

In the proof graph of Fig. 73.2, the action $g(v, w)$ decreases dev_r; and no other action would affect dev_l or dev_r. Hence, each node may be extended by the requirement $dev_l < n \wedge dev_r \leq m$, which yields

$$(5) \wedge dev_l < n \wedge dev_r \leq m \hookrightarrow ground \wedge dev_l < n \wedge dev_r \leq m. \tag{19}$$

Likewise follows with the proof graph of Fig. 73.3:

$$(14) \wedge dev_l \leq n \wedge dev_r < m \hookrightarrow ground \wedge dev_l \leq n \wedge dev_r < m. \tag{20}$$

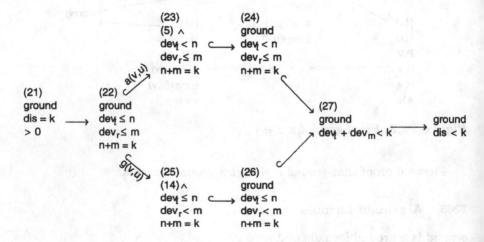

Figure 73.4. Proof graph for $\Sigma_{73.1}$

We are now prepared to justify the nodes of the proof graph in Fig. 73.3:

(21) by (18)

(22) inv2 prevents e, and inv4 prevents c, pick-up pattern of Sect. 69.5
(23) by (19)
(24) by propositional reasoning
(25) by (20)
(26) by propositional reasoning
(27) by (18)

To each *ground*-state there exists an index k with $dis = k$. Then finitely many instantiations of the proof graph of Fig. 73.1 yields

$$ground \hookrightarrow ground \land dis = 0. \tag{28}$$

With the proposition of Sect. 73.3 and Theorem 70.2 follows

$$s_{\Sigma_{73.1}} \hookrightarrow ground. \tag{29}$$

Hence (28) and (29) together with (4) give

$$s_{\Sigma_{73.1}} \hookrightarrow rearr, \tag{30}$$

which describes, as intended, that rearrangement will be reached inevitably.

74   Self-Stabilizing Mutual Exclusion

74.1   Properties to be shown

Figure 74.1 recalls algorithm $\Sigma_{26.2}$, renaming its places. Assuming a concrete

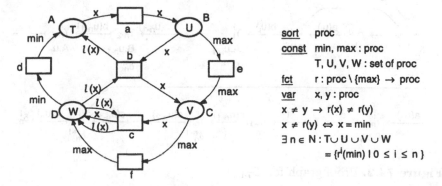

Figure 74.1. Renamed self-stabilizing mutex $\Sigma_{26.2}$

interpretation of the involved constant symbols, let $R = \{u_1, \ldots, u_n\}$ be the set of processes (i.e., $R = T \cup U \cup V \cup W$, with $u_1 = min$, $u_n = max$, and $r(u_i) = u_{i+1}$ $(i = 1, \ldots, n)$).

A state is feasible if it fulfills the equation

$$A + B + C + D = R. \tag{1}$$

Two properties of $\Sigma_{74.1}$ are to be shown: Firstly, each feasible state leads to a state with all processes at D:

$$\text{feasible} \mapsto D.R, \tag{2}$$

and secondly, for each state reachable from a $D.R$-state holds:

$$|A| \le 1. \tag{3}$$

74.2   Proof of (2)

For each non-$D.R$-state s, call i the *smooth index* of s if $s \models \neg D.u_i$ and $s \models D.u_{i+1}, \ldots, u_n$. By definition let 0 be the smooth index of the state $D.R$. The proof graph

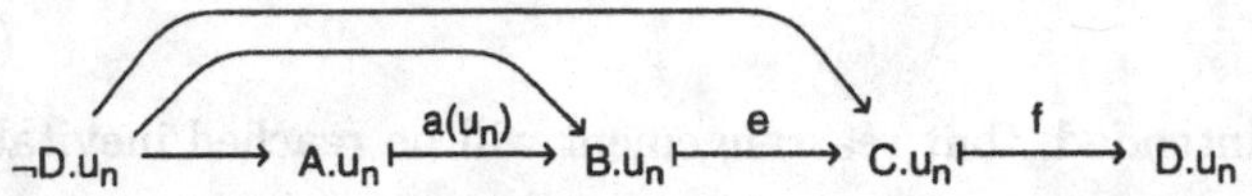

$$\tag{4}$$

shows that each feasible state with smooth index n leads to a state with a smaller smooth index.

Inductively, we show that each state with smooth index i leads to a state with a smaller smooth index. To this end we introduce shorthands $\gamma_j^i :=$ $C.u_i \wedge \ldots \wedge C.u_j$ and $\delta_j^i := D.u_i \wedge \ldots \wedge D.u_j$. Figure 74.2 then shows the required property. A $D.R$-state will be reached after at most n iterations.

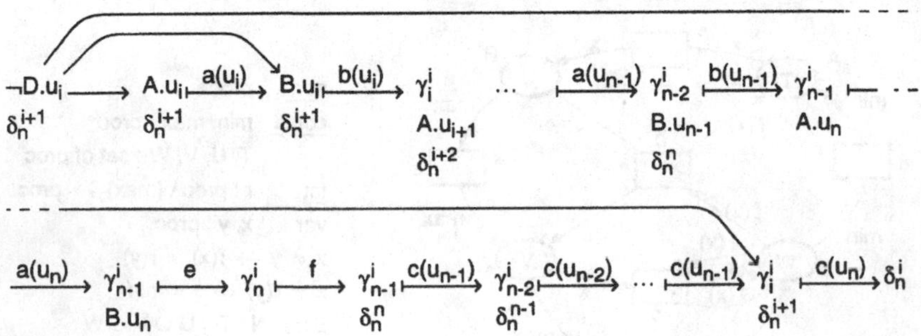

Figure 74.2. Proof graph for $\Sigma_{74.1}$

74.3   Proof of (3)

Here we assume $D.R$ as the initial state of $\Sigma_{74.1}$. Then for each $u \in R$, $u \neq$ min, the set $\{C.u, D.r(u)\}$ is a trap. Hence for all $u \neq$ min, $C.u + D.r(u) \ge 1$ (by Theorem 61.2), hence

$$|C| + |D| \geq n - 1. \tag{5}$$

Furthermore, (1) implies

$$|A| + |B| + |C| + |D| = n. \tag{6}$$

Then the inequality $(6)-(5)= |A| + |B| \leq 1$ immediately yields (3).

75 Master/Slave Agreement

75.1 The essential property

Figure 75.1 recalls the master/slave algorithm $\Sigma_{30.1}$, renaming its places. The essential aspect of $\Sigma_{30.1}$ is to guarantee that *master pending* is eventually followed by *master inactive* together with either all slaves *busy* or all slaves *pending*. In the redrawn version $\Sigma_{75.1}$ of $\Sigma_{30.1}$ this property is formally represented by

$$\Sigma_{75.1} \models B \hookrightarrow A \land (N.U \lor P.U). \tag{1}$$

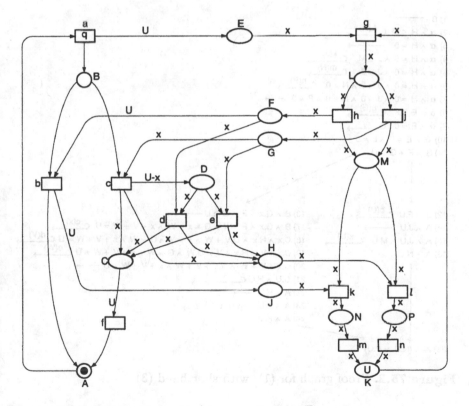

Figure 75.1. Renamed master/slave agreement $\Sigma_{30.1}$

75.2   State properties

Proof of (1) is based on the following place invariants of $\Sigma_{75.1}$:

> inv1: $E + L + F + G - D - U * |B| = 0$
> inv2: $F + G + H + J + N + P + K + L = U$
> inv3: $U * A + U * B + C + D = U$
> inv4: $F + G + J + H - M = 0$
> inv5: $H + J + N + P + K - E - U * A - C = 0$
> inv6: $L + M + N + P + K = U$

inv4 and inv6 imply $F + G + J + H \leq U$.   (2)

75.3   A proof graph for the essential property

Figure 75.2 shows a proof graph for (1). As a shorthand it employs

$$\alpha = B \wedge (E + L + F + G \geq U).$$ (3)

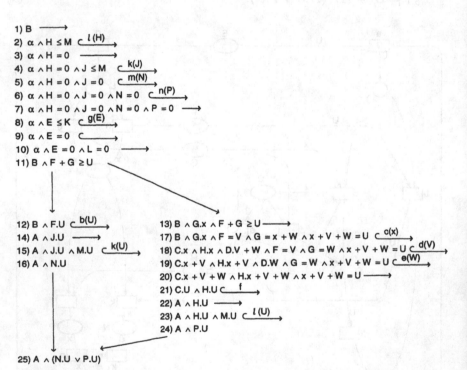

Figure 75.2. Proof graph for (1), with shorthand (3)

Four sections can be distinguished in this proof graph: The steps from line 2 to line 7 finish the previous round. Line 7 to line 11 start the actual round, proceeding until each slave has made its choice. The left branch describes the case of all sites agreed, with occurrence of $b(U)$ and $k(U)$. The right branch describes the case of at least one slave refusing, with occurrence of c, d, e, and, most important, $l(U)$.

The nodes of this proof graph are justified as follows, with rule numbers referring to Sect. 69:

1. inv1, inv4.
2. $H.x \rightarrow \neg J.x$ by (2), hence $H.x$ prevents $k(x)$, hence the proposition with pattern of Sect. 69.4.
3. inv4.
4. $J.x \rightarrow \neg H.x$ by inv7, hence $J.x$ prevents $l(x)$, hence the proposition with pattern of Sect. 69.4.
5. pattern of Sect. 69.1.
6. pattern of Sect. 69.1.
7. inv5.
8. pattern of Sect. 69.3.
9. pattern of Sect. 69.2.
10. definition of α.
11. propositional logic, inv2.
12. $F.U \rightarrow G = 0$ by (2), hence $F.U$ prevents $c(x)$. Furthermore, $B \rightarrow D = 0$ by inv3, hence B prevents d. Hence the proposition with pattern of Sect. 69.5.
14. inv4.
15. $J.U \rightarrow H = 0$ by (2), hence $J.U$ prevents $l(x)$. Furthermore, $J \leq M$, with inv4. Hence the proposition with pattern of Sect. 69.4.
16. propositional logic.
13. inv2.
17. $G.x \rightarrow \neg F.U$ by (2), hence $G.x$ prevents b. Furthermore, $B \rightarrow D = 0$ by inv3, hence B prevents $e(x)$. Hence the proposition with pattern of Sect. 69.5.
18. The case of $V + W = 0$ directly implies 21. Otherwise $C < U$ by inv6, which prevents f. Furthermore, $C.x \rightarrow \neg B$ by inv3, hence $C.x$ prevents b. Finally, $F.V \rightarrow G \cap V = \emptyset$ by inv2, hence $F.V$ prevents $e(y)$ for each $y \in V$. Hence the proposition with pattern of Sect. 69.4 and context $H.x \wedge D.W \wedge G = W$.
19. The case of $W = 0$ directly implies 21. Otherwise $C < U$ by inv6, which prevents f. Furthermore, $C.x \rightarrow \neg B$ by inv3, hence $C.x$ prevents $c(y)$ for each $y \in W$. Finally, $G.W \rightarrow F \cap W = \emptyset$ by inv2, hence $G.W$ prevents $d(y)$ for each $y \in W$. Hence the proposition with pattern of Sect. 69.4 and context $C.V \wedge H.x + V$.
20. propositional implication.
21. pattern of Sect. 69.1 with context $H.U$.
22. inv4.

23. $H.U \to J = 0$ by (2), hence $H.U$ prevents $k(x)$. Furthermore, $H \leq M$ with inv4. Hence the proposition with pattern of Sect. 69.4.

24. propositional logic.

76   Leader Election

Figure 76.1 recalls $\Sigma_{32.1}$ with renamed places. With max the maximal element of U, the property to be shown is

$$s_{\Sigma_{76.1}} \hookrightarrow B.U \times \{\text{max}\} \wedge A = \emptyset \wedge C = \emptyset. \tag{1}$$

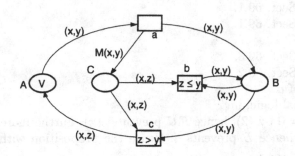

sort	site		var	x, y, z : site
sort	state : site × site			$x,y \in U \to x\,W^* y$
const	U : set of sites			$W_1 \cup W_2 = U$
const	V,W : set of states			$V = \{(u,u) \mid u \in U\}$
	≤ : total order on U			$M(x,y) = W(x) \times \{y\}$
fct	M : state → set of states			

Figure 76.1. Renamed leader election $\Sigma_{32.1}$

76.1   Fundamental state properties

An obvious place invariant implies that each site is either *pending* or *updating*:

$$A_1 + B_1 = U. \tag{2}$$

Furthermore, a site v, already knowing the leader, is related to its neighbors by a property derived from a trap. To this end, assume a state s and two neighboring sites $u, v \in U$, and $s \models B.(u, \text{max})$. s has been reached by occurrence of $a(u, \text{max})$. This action also produced $C.(v, \text{max})$. With s considered as (a new) initial state, an initialized trap yields the inequality $A.(u, \text{max}) + C.(v, \text{max}) + A.(v, \text{max}) + B.(v, \text{max}) \geq 1$. Together with (2) this yields the valid propositional formula

$$B.(u, \max) \lor C.(v, \max) \lor A.(v, \max) \lor B.(v, \max). \tag{3}$$

Intuitively formulated, each neighbor of a site already updating with the leader is also aware of the leader, or a corresponding message is pending.

76.2   A fundamental progress property

A weight function f will be required, that assigns each state (u, v) its "better" candidates. So, for all $u, v \in U$ let

$$f(u, v) = \{(u, w) \mid w > v\}. \tag{4}$$

Obviously, $f(u, v) = \emptyset$ if $v = \max$.

We stick to states with all sites updating $(B_1.U)$ in the sequel. This includes the terminal state with no pending messages $(C = \emptyset)$ and empty weights $f(u, v)$ for all sites u $(f(B) = \emptyset)$. The proof graph of Fig. 76.2 states

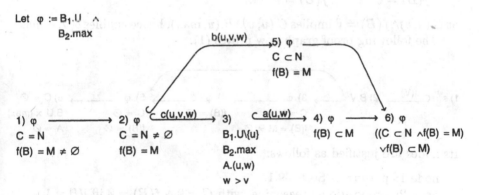

Figure 76.2. Proof graph for $\Sigma_{76.1}$

that one of the sites not yet knowing the leader $(f(B) = M \neq \emptyset)$ will eventually hold a "better" candidate $(f(B) \subset M)$, or will have skipped a pending message $(C \subset N)$. The proof graph's nodes are justified as follows:

node 1: $B_2.\max$, $f(B) \neq \emptyset$ and the graph's connectedness imply neighboring sites u and w, $B.(u, \max)$, and $B.(v, i)$ with $i < \max$. Then $C.(u, \max)$ by (3) and (2).

node 2: $C \neq \emptyset$ implies some $C.(u, w)$, and φ implies some $B.(u, v)$. This enables $b(u, v, w)$ or $c(u, v, w)$. Hence, $C \neq \emptyset \land \varphi$ is progress prone. Then apply the pattern of Sect. 69.5.

node 3:   pattern of Sect. 69.2.

nodes 4 and 5: propositional logic.

76.3 Proof of (1)

The proof graph in Fig. 76.2 shows

$$\begin{array}{cc}\varphi & \varphi \\ C = N & \overset{\hookrightarrow}{} \quad ((C \subset N \wedge f(B) = M) \quad . \\ f(B) = M \neq \emptyset & \vee f(B) \subset M) \end{array} \tag{5}$$

C may shrink finitely often only, hence finitely many iterations of (5) yield

$$\begin{array}{cc}\varphi & \varphi \\ C = N & \overset{\hookrightarrow}{} \quad f(B) \subset M \quad . \\ f(B) = M \neq \emptyset & \end{array} \tag{6}$$

A remaining message is cleared by

$$\begin{array}{cc}\varphi & \varphi \\ C = N & \overset{\hookrightarrow}{} \quad C \subset N \quad , \\ f(B) = \emptyset & f(B) = \emptyset \end{array} \tag{7}$$

as $C.(u, v) \wedge f(B) = \emptyset$ implies $C.(u, v) \wedge B.(u, \max)$, hence enables $b(u, v, \max)$. The following proof graph now proves (1):

$$\begin{array}{cccccc} 1) \ s_\Sigma & \overset{a(V)}{\longrightarrow} & 2) \ B.V & \longrightarrow & 3) \ \varphi & \overset{C}{\longrightarrow} & 4) \ \varphi & \overset{C}{\longrightarrow} & 5) \ \varphi & \longrightarrow & 6) \ C = \emptyset \\ & & & & C = N & & f(B) = \emptyset & & C = \emptyset & & B.U \times \{\max\} \\ & & & & f(B) = M \neq \emptyset & & & & f(B) = \emptyset & & A = \emptyset \end{array}$$

Its nodes are justified as follows:

node 1: pattern of Sect. 69.1.
node 2: propositional reasoning, with $C = \emptyset \wedge f(B) = \emptyset$ iff $|U| = 1$.
node 3: finitely many iterations of (6).
node 4: finitely many iterations of (7).
node 5: by construction of φ.

77   The Echo Algorithm

77.1 Properties to be proven

Figure 77.1 provides a redrawn version of the Echo Algorithm of Fig. 33.2. It has two decisive properties: Firstly, the initiator terminates only if all other sites have been informed before. In Fig. 77.1, this reads

$$C.i \to G.U \tag{1}$$

and is a typical state property. Secondly, the initiator *will* eventually terminate, i.e.,

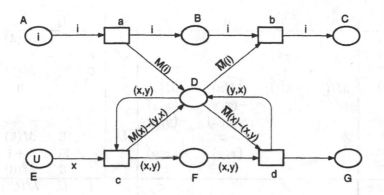

sort	site		$W = W^{-1}$
sort	message = site × site		$x,y \in U \cup \{i\} \rightarrow x\, W^* y$
const	i : site		$W_1 = U \cup \{i\}$
const	U : set of sites		$i \notin U$
const	W : set of (sites × sites)		$M(x) = W(x) \times \{x\}$
fct	M, $\overline{M}$: site → set of messages		$\overline{M}(x) = M(x)^{-1}$
var	x,y : site		

Figure 77.1. Redrawn echo algorithm $\Sigma_{33.2}$

$$s_{\Sigma_{77.1}} \mapsto C.i, \tag{2}$$

which is a typical liveness property. Both (1) and (2) will be verified in the sequel.

There is no straightforward place invariant or trap that would prove (1). Nor is there a proof graph for (2), with steps picked up according to the patterns of Sect. 69. Rather, one has to argue inductively along a spanning tree that yields at place F.

77.2   Three place invariants

Figure 77.1 has three important place invariants, as given in Fig. 77.2. Two of them are inductively quite obvious, representing the "life lines" of the initiator i and of all other sites, respectively.

The equation of I_1 is $A + B + C = i$. This implies

$$A.i + B.i + C.i = 1, \tag{3}$$

hence the initiator is always either at its start or is waiting, or is terminated. The equation furthermore implies

$$\forall x \in U : A.x + B.x + C.x = 0, \tag{4}$$

hence no non-initiator site ever finds at A, B, or C.

Correspondingly, the equation of I_2 is $E + F_1 + G = U$. This implies

	a	b	c	d	s_Σ	I_1	I_2	I_3
A	$-i$				i	A		$M(A)$
B	i	$-i$				B		
C		i				C		$\overline{M}(C)$
D	$M(i)$	$-\overline{M}(i)$	$M(x)$ $-(x,y)$ $-(y,x)$	$-\overline{M}(x)$ $+(x,y)$ $+(y,x)$				D
E			$-x$		U		E	$M(E)$
F			(x,y)	$-(x,y)$			F_1	$F+\overline{F}$
G				x			G	$\overline{M}(G)$
$I \cdot s_\Sigma$						i	U	$M(U')$

Let $\overline{F} = F^{-1}$ and $U' = U \cup \{I\}$

Figure 77.2. Matrix, initial state, and three place invariants of $\Sigma_{77.1}$

$$\forall x \in U : E.x + F_1.x + G.x = 1, \tag{5}$$

hence each non-initiator is always either uninformed or pending or informed. The equation furthermore implies

$$\forall x \notin U : E.x + F_1.x + G.x = 0, \tag{6}$$

hence the initiator never finds on E, F, or G.

I_3, finally, represents the *potential messages* of the system. Its equation is $M(A) + \overline{M}(C) + D + M(E) + F + \overline{F} + \overline{M}(G) = M(U')$, implying for each message $(y,x) \in M(U')$ the property $M(A).(y,x) + \overline{M}(C).(y,x) + D.(y,x) + M(E).(y,x) + F.(y,x) + \overline{F}.(y,x) + \overline{M}(G).(y,x) = M.(y,x)$, which in turn reduces to

$$\begin{aligned} \forall x \in U' \quad \forall y \in W(x) : \\ A.x + C.y + D.(y,x) + E.x + F.(y,x) + F.(x,y) + G.y = 1. \end{aligned} \tag{7}$$

Hence for each message (y,x) holds: Its sender x is still starting or uninformed, or the message has already been sent but not received yet, or one of y and x has received the message from x to y, respectively, or the message's receiver y is terminated or informed.

77.3   The pending site's rooted tree

A further state property will be required, stating that the tokens on F always form a tree with root i. This will be formulated with the help of the following notation:

A sequence $u_0 \ldots u_n$ of sites $u_i \in U'$ is a *sequence of F at a state* s iff $s \models F.(u_{i-1}, u_i)$ for $i = 1, \ldots, n$. $\tag{8}$

For each reachable state s we will now prove the following two properties:

For each $F_1.u$ there is a unique sequence $u_0 \ldots u_n$ of F with $u_0 = u$ and $u_n = i$, $\tag{9}$

and

the elements of each sequence of F are pairwise different.   (10)

Both properties now are together shown by induction on the reachability of states:

Both (9) and (10) hold initially, as $s_{\Sigma_{77.1}} \models F = \emptyset$. Now, let r be a reachable state, let $r \xrightarrow{m} s$ be a step of some transition t, and inductively assume (9) and (10) for r.

The case of $t = a$ or $t = b$ implies $r(F) = s(F)$, hence the step $r \xrightarrow{m} s$ retains both (9) and (10) for s. For $t = c$ or $t = d$ let $m(x) = u$ and $m(y) = v$.

The case of $t = c$ goes as follows: Enabledness of $c(m)$ at r now for r implies $D.(u, v)$ and $E.u$. Then $r \models F_1.v$, according to the following sequence of implications:

$$
\begin{array}{ccccc}
1. \longrightarrow & 2. \longrightarrow & 3. \longrightarrow & 4. \longrightarrow & 5. \\
D.(u,v) & D.(u,v) & \neg E.v & E.v & F_1.v \\
E.u & E.u & E.u & \neg G.v & \\
 & v \in W(u) & v \in W(u) & &
\end{array}
$$

Its nodes are justified as follows:

    node 1: (6);
    node 2: (7) with $x = v$, $y = u$;
    node 3: (7) with $x = u$, $y = v$;
    node 4: (5).

Now, $r \models F_1.v$ and the inductive assumption of (9) imply a unique sequence $v \ldots i$ of F at state r. Then $uv \ldots i$ is a sequence of F at state s, because $s(F) = r(F) + (u, v)$. Together with (5), this implies (9) for s. Furthermore, $r \models u \notin F_1$ (by (5)) and $u \neq i$ by (4), hence (10) for s.

Correspondingly, enabledness of $d(m)$ at r now for r implies $D.\overline{M}(u) - (u, v)$ and $F.(u, v)$. Then $r \models F_2.u$ according to the following sequence of implications:

$$
\begin{array}{cccccc}
1. \longrightarrow & 2. \longrightarrow & 3. \longrightarrow & 4. \longrightarrow & 5. \longrightarrow & 6. \\
D.\overline{M}(u) & D.\overline{M}(u) & F \cap (\overline{M}(u) & F \cap (M(u) & F \cap M(u) = \emptyset & \neg F_2.u \\
-(u, v) & -(u, v) & -(u, v)) = \emptyset & -(v, u)) = \emptyset & & \\
F.(u, v) & \neg F.(v, u) & \neg F.(v, u) & \neg F.(v, u) & &
\end{array}
$$

Its nodes 1 and 2 are justified by (7), nodes 3, 4, and 5 by properties of M.

With $r \models \neg F_2.u$, for each sequence $u_0 \ldots u_n$ of F, $u_1, \ldots, u_n \neq u$. This implies (9) for the state s, because $s(F) = r(F) - (u, v)$. (10) is then trivial, because $s(F) \subseteq r(F)$.

77.4 Proof of the state property (1)

(1) is indirectly proven in three steps:

i. Assume $F \neq \emptyset$. Then there exists some $w \in U'$ with $F.(w, i)$, by (9). Then $\neg C.i$ by (7).

ii. For all $u \in U'$ we show

$$E.u \rightarrow \neg C.i \qquad (*)$$

by induction on the distance of u to i: For $u = i$, (*) holds trivially, as $\neg E.i$ by (6). Inductively assume (*), let $v \in W(u)$, and assume $E.v$. Then $u \in W(v)$, hence $\neg G.u$, by (7). Then $F_1.u$ or $E.u$, by (5). The case of $F_1.u$ implies $F \neq \emptyset$, hence $\neg C.i$ by (i). The case of $E.u$ implies $\neg C.i$ by inductive assumption.

iii. $C.i \rightarrow E = F = \emptyset$, by (i) and (ii). Then (1) follows from (5).

77.5 Progress from *uninformed* to *pending*

Here we show that each *uninformed* site $u \in U$ will eventually go *pending*. In terms of $\Sigma_{77.1}$ this reads:

Let $U = V \cup W$, $V \neq \emptyset$, $W \neq \emptyset$. Then

$$E.V \wedge F_1.W \hookrightarrow \bigvee_{v \in V} (E.V - v \wedge F_1.W + v). \qquad (11)$$

This property holds due to the following proof graph:

1) $E.V \wedge F_1.W \wedge V \neq \emptyset \wedge W \neq \emptyset \rightarrow$
2) $E.V \wedge F_1.W \wedge \text{ex. } v \in V \wedge \text{ex. } w \in W \cup \{i\}$ with $D.(v, w) \hookrightarrow$
3) $E.V - v \wedge F_1.W + v$

Its nodes are justified as follows:

node 1: Connectedness of U' implies some neighbors v, w such that $E.v$, and $F_1.w$ or $w = i$. Furthermore,

    i. $F_1.w$ implies $w \in U$ by (6), hence $\neg A.w$ by (4). $w = i$ and $W \neq \emptyset$ imply some $F.(u, i)$ by (9), hence $\neg A.i$ by (7).

    ii. $E.v$ implies $v \in U$ by (6), then $\neg C.v$ by (4).

    iii. $F_1.w$ implies $\neg E.w$ by (5) and $w = i$ implies $\neg E.w$ by (6).

    iv. $E.v$ implies $\neg F_1.v$ by (9), hence $\neg F.(v, w)$.

    v. Let $u_0 \ldots u_n$ be a sequence of F with $u_0 = w$ and $u_n = i$, according to (9). The case of $n = 1$ implies $u_1 = i \neq v$, hence $\neg F.(w, v)$. Otherwise, $F_1.u_1$. Then $E.v$ implies $u_1 \neq v$ by (5). Hence $\neg F.(w, v)$.

    vi. $E.v$ implies $\neg G.v$ by (5).

    Now (i),...,(vi), and (7) imply $D.(v, w)$.

node 2: pattern of Sect. 69.5.

77.6 Progress from *pending* to *informed*

Here we show that each pending site will eventually be informed. In terms of $\Sigma_{77.1}$ this reads:

Let $U = V \cup W$ with $V \neq \emptyset$. Then

$$F_1.V \wedge G.W \hookrightarrow \bigvee_{v \in V}(F_1.V - v \wedge G.W + v). \tag{12}$$

This property holds due to the following proof graph:

1) $F_1.V \wedge G.W \wedge V \cup W = U \wedge V \neq \emptyset \to$
2) ex. $v \in V$ ex. $w \in U$:
   $F_1.V \wedge G.W \wedge V \cup W = U \wedge D.(\overline{M}(v) - (v,w)) \hookrightarrow$
3) ex. $v \in V$ ex. $w \in U$ with $F_1.V - w \wedge G.W + v$.

Its nodes are justified as follows:

node 1: Let $u_0 \ldots u_n$ be a maximal sequence of F. This exists due to
(9) and (10). In case u_1 is the only neighbor of u_0, $D.(\overline{M}(u_0) - (u_0, u_1)) = D.((u_0, u_1) - (u_0, u_1)) = D.\emptyset$ which holds trivially.
Otherwise, let $(u_0, v) \in \overline{M}(u_0) - (u_0, u_1)$. Then the following six
properties hold:
   i. (9) implies some $F.(w, i)$, hence $\neg A.i$ by (7), hence $\neg A.v$ in
      case $i = v$. Otherwise, $v \in U$, hence $\neg A.v$ by (4).
   ii. $u_0 \in U$ by construction, hence $\neg C.u_0$ by (4).
   iii. $E = \emptyset$ by (5) and $V \cup W = U$, hence $\neg E.v$.
   iv. Maximality of $u_0 \ldots u_n$ implies $\neg F.(v, u_0)$.
   v. $F.(u_0, u_1)$ implies $\neg F.(u_0, v)$ as the path from u_0 to i is unique
      by (9).
   vi. $F_1.u_0$ implies $\neg G.u_0$.
   Now (i),$\ldots$,(vi), and (7) imply $D.(u_0, v)$. This argument applies
   to all $(u_0, v) \in \overline{M}(u_0) - (u_0, u_1)$, hence $D.\overline{M}(u_0) - (u_0, u_1)$.
node 2: pattern of Sect. 69.4.

77.7 Proof of the liveness property (2)

(2) is now proven with the help of the proof graph of Fig. 77.3. Its nodes are
justified as follows:

node 1: definition of a_Σ
node 2: pattern of Sect. 69.1, context E
node 3: $c.(u, i)$ is enabled for each $u \in M(i)$; pattern of Sect. 69.4
node 4: $|V|$-fold application of (11)
node 5: $|U|$-fold application of (12)
node 6: we distinguish three cases:
   i. $u \in M(i)$ implies $u \neq i$, hence $\neg A.u$ by (4)
   ii. $G.U$ implies $E = F = \emptyset$ by (5) and (6). Hence $\neg E.u$,
       $\neg F.(i, u)$, and $\neg F.(u, i)$.

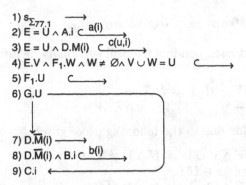

Figure 77.3. A proof graph for $s_{\Sigma_{77.1}} \hookrightarrow C.i$

    iii. $i \notin U$ implies $\neg G.i$ by (6).

    Now, (i), (ii), and (iii) with (7) imply $D.(i,u) \vee C.i$. This argument
applies to all $(i,u) \in \overline{M}(i)$, hence $D.\overline{M}(i) \vee C.i$.

node 7: $\neg C.i$ by (7); $\neg A.i$ because $s_{\Sigma_{77.1}} \to \neg D.\overline{M}(i)$, the only initial
    step is $s_{\Sigma_{77.1}} \xrightarrow{a} B.i$, and $\{B.i, C.i\}$ is a trap, initialized after this
    step. Hence the proposition by (3).

node 8: pattern of Sect. 69.1

78   Global Mutual Exclusion on Undirected Trees

78.1   The property to be proven

Here we consider the version of Fig. 34.2. There is one progress property
to be shown for this algorithm: each request of a site u for going *critical*
(i.e., $job.(u,y)$) is eventually served (i.e., *critical.u*). In terms of the redrawn
version as in Fig. 78.1, this reads

$$B.(x,x) \mapsto D.x. \tag{1}$$

Proof of (1) is based on state properties, to be considered first.

78.2   State properties

The forthcoming two state properties exploit the cycle free structure of the
underlying network. A basic state property of $\Sigma_{78.1}$ is a tree on $\overline{E} \cup B \cup C$
with its root in G (where $\overline{E}$ stands for E^{-1}). More precisely,

> With $G.u$, the tokens on $\overline{E} \cup B \cup C$ consist of paths from u to
> all sites v. Those paths form a tree. Tokens formed (u,u) may   (2)
> additionally occur at B or C.

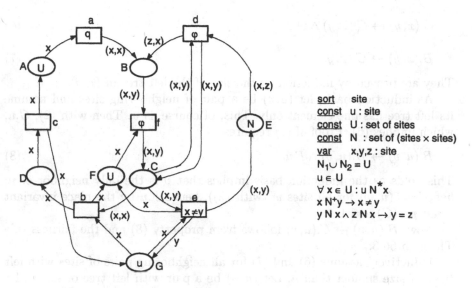

Figure 78.1. Renamed global mutex on trees, $\Sigma_{34.2}$

This property can easily be proven by induction on the reachability of states: (2) obviously holds initially, due to the assumption given in Fig. 78.1. Occurrences of d, f, or e in any mode apparently preserve the tree on $\overline{E} \cup B \cup C$. Occurrences of a and b add and remove pairs formed (u, u) to and from C, respectively. Action c, finally, does not touch E, B, or C.

A further state property is based on the *left tree*, defined for each pair (u, v) of neighboring sites u and v: Any site w belongs to the left tree of (u, v) if in the underlying network, the undirected path connecting w and u does not include v. The following property of left trees will be exploited in inductive proofs:

> If w and v are neighbors of u, then the left tree of (w, u) is smaller than the left tree of (u, v). $\hfill (3)$

Apparently, we have for $G.r$ and each site u:

> If $r \neq u$, there is a unique neighbor site w of u, with r in the left tree of (w, u). $\hfill (4)$

Then with (2),

> if $E.(v, u)$ or $B.(u, v)$ or $C.(u, v)$, and if $G.w$, then w is in the left tree of (u, v). $\hfill (5)$

78.3 Progress properties

Two properties of pairs (u, v) of neighboring sites u and v are considered here:

$$C.(x,y) \mapsto C.(x,y) \wedge G.x \tag{6}$$

$$B.(x,y) \mapsto C.(x,y) \tag{7}$$

They are proven by induction on the size of the left tree of (x,y).

As induction basis, let (u,v) be a pair of neighboring sites and assume its left tree has one element only. This, of course, is u. Then with (2), $G.u$, which implies (6). Proof of (7) requires

$$B.(u,v) \to B.(u,v) \wedge F.u. \tag{8}$$

This holds as the induction basis implies that v is the only neighbor of u; hence $\neg C.(u,w)$ for all sites w (with (2)), hence $F.u$ by the place invariant $F.u + pr_1(C).u = 1$.

Now, $B.(u,v) \mapsto C.(u,v)$ follows from property (8) and the fairness rule Theorem 66.3.

Inductively assume (6) and (7) for all neighboring pairs of sites with left trees of size smaller than n. Let (u,v) be a pair with left tree of size n, let $G.r$, and let w be the neighbor of u, constructed according to (4).

Then the following proof graph proves (6) for $(x,y) = (u,v)$:

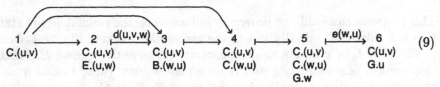

$$\tag{9}$$

Its nodes are justified as follows:

node 1: by (4) and (2).
node 2: fairness rule Theorem 66.3.
node 3: inductive assumption of (7) for $(x,y) = (w,u)$, and (3).
node 4: inductive assumption of (6) for $(x,y) = (w,u)$, and (3).
node 5: pick-up rule of Sect. 69.3.

Proof of (7) requires the proof graph

$$1) B.(u,v) \longrightarrow 2) C.(u,v') \longmapsto 3) \begin{array}{c} C.(u,v') \\ G.u \end{array} \xrightarrow{e(u,v')} 4) F.u \tag{10}$$

Its nodes are justified as follows:

node 1: place invariant $C_2.u + F.u = 1$.
node 2: $v' \neq w$ by (2); then w constructed according to (4) for v and for v' coincide; then (9) with v replaced by v'.
node 3: pick-up rule of Sect. 69.3.

Now, $B.(u,v) \mapsto C.(u,v)$ follows with property (10) and the fairness rule Theorem 66.3.

78.4 Proof of (1)

Two properties are required, provided by the following two proof graphs:

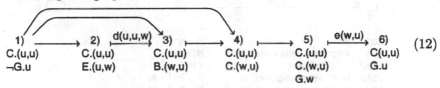

1) B.(u,u) ⟶ 2) C.(u,v) ⊢⟶ 3) C.(u,v) ∧ G.u ⊢$\xrightarrow{e(u,v)}$ 4) F.u   (11)

node 1: place invariant $C_2.u + F.u = 1$.
node 2: property (6).
node 3: pick-up rule of Sect. 69.2.

The second proof graph is

$$
\begin{array}{ccccccc}
1) & \xrightarrow{} & 2) & \xrightarrow{d(u,u,w)} & 3) & \vdash\!\!\longrightarrow & 4) & \vdash\!\!\longrightarrow & 5) & \vdash\!\!\xrightarrow{e(w,u)} & 6) \\
C.(u,u) & & C.(u,u) & & C.(u,u) & & C.(u,u) & & C.(u,u) & & C.(u,u) \\
\neg G.u & & E.(u,w) & & B.(w,u) & & C.(w,u) & & C.(w,u) & & G.u \\
& & & & & & & & G.w & &
\end{array}
$$
(12)

node 1: Let w be the neighbor of u, constructed according to (4). Then
apply (2).
node 2: fairness rule Theorem 66.3.
node 3: property (7).
node 4: property (6).
node 5: pick-up rule of Sect. 69.3.

Now, (1) is proven by the following proof graph for $x = u$:

$$
\begin{array}{ccccccccc}
1) & \vdash\!\!\longrightarrow & 2) & \vdash\!\!\longrightarrow & 3) & \vdash\!\!\longrightarrow & 4) & \vdash\!\!\longrightarrow & 5) \\
B.(u,u) & & C.(u,u) & & C.(u,u) & & C.(u,u) & & D.u \\
& & & & \neg G.u & & G.u & &
\end{array}
$$
(13)

with

node 1: property (11), fairness rule Theorem 66.3.
node 2: propositional argument.
node 3: property (12).
node 4: pick-up rule of Sect. 69.3.

79 Local Mutual Exclusion

79.1 Properties to be proven

As in most cases, safety and liveness properties are to be proven. A safety
property guarantees that neighboring sites are never both *critical* at the same
time. In terms of the redrawn representation of Fig. 79.1 this reads

$$D.x \wedge y \in r(x) \rightarrow \neg D.y. \tag{1}$$

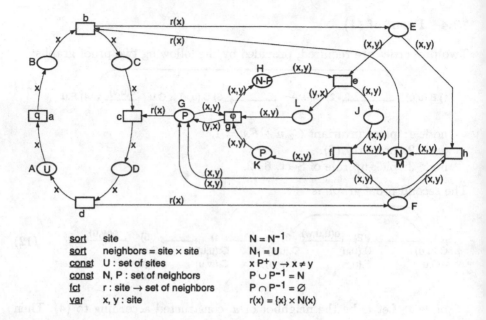

sort	site		$N = N^{-1}$
sort	neighbors = site × site		$N_1 = U$
const	U : set of sites		$x\,P^+\,y \rightarrow x \neq y$
const	N, P : set of neighbors		$P \cup P^{-1} = N$
fct	r : site → set of neighbors		$P \cap P^{-1} = \varnothing$
var	x, y : site		$r(x) = \{x\} \times N(x)$

Figure 79.1. Renamed mutex on networks $\Sigma_{34.3}$

A liveness property guarantees *evolution*: Each *pending* site is eventually *critical*:

$$B.x \mapsto D.x. \tag{2}$$

Proof of (1) and (2) starts with some state properties of $\Sigma_{79.1}$, derived from place invariants or from inductive arguments. There is always a distinguished *partial order* on the sites, describing *priority* of access to shared resources. Upon going *critical*, a site may have priority for some, but not necessarily all resources.

79.2   State properties

We start with eleven state properties from place invariants of $\Sigma_{79.1}$:

$A + B + C + D = U$, hence for each $u \in U$

$$A.u + B.u + C.u + D.u = 1. \tag{3}$$

$r(C) + r(D) + F + M = N$, hence for each $(u, v) \in N$

$$C.u + D.u + F.(u, v) + M.(u, v) = 1. \tag{4}$$

$r(C) + r(D) + F - J - E = 0$, hence for each $(u, v) \in N$

$$C.u + D.u + F.(u, v) - J.(u, v) - E.(u, v) = 0. \tag{5}$$

$r(C) + r(D) + F + K + H - E = N$, hence for each $(u, v) \in N$

$$C.u + D.u + F.(u,v) + K.(u,v) + H.(u,v) - E.(u,v) = 1. \tag{6}$$

$E + J + M = N$, hence for each $(u,v) \in N$

$$E.(u,v) + J.(u,v) + M.(u,v) = 1. \tag{7}$$

$H + J + K = N$, hence for each $(u,v) \in N$

$$H.(u,v) + J.(u,v) + K.(u,v) = 1. \tag{8}$$

$G + \overline{G} + r(D) + r(\overline{D}) + F + \overline{F} = P + \overline{P} = N$, hence for each $(u,v) \in N$

$$G.(u,v) + G.(v,u) + D.u + D.v + F.(u,v) + F.(v,u) = 1. \tag{9}$$

$G + r(D) + F + H + \overline{L} = N$, hence for each $(u,v) \in N$

$$G.(u,v) + D.u + F.(u,v) + H.(u,v) + L.(v,u) = 1. \tag{10}$$

$L + \overline{L} + \overline{H} - J - K = 0$, hence for each $(v,u) \in N$

$$L.(v,u) + L.(u,v) + H.(u,v) - J.(v,u) - K.(v,u) = 0. \tag{11}$$

$r(C) - J - E - G - H - \overline{L} = N$, hence for each $(v,u) \in N$

$$C.v - J.(v,u) - E.(v,u) - G.(v,u) - H.(v,u) - L.(u,v) = 1. \tag{12}$$

$r(C) + K - E - G - \overline{L} = 0$, hence for each $(u,v) \in N$

$$C.u + K.(u,v) - E.(u,v) - G.(u,v) - L.(v,u) = 0. \tag{13}$$

In addition, the following two properties will be required:

$$K.(u,v) \rightarrow G.(u,v) \vee D.u \vee F.(u,v). \tag{14}$$

This property holds initially and is apparently preserved by occurrences of $c(u)$ and $d(u)$. Occurrence of $f(u,v)$ or $h(u,v)$ lead to both $K.(u,v)$ and $G.(u,v)$. Occurrence of $g(u,v)$ leads to both $\neg G.(u,v)$ and $\neg K.(u,v)$. No other occurrences of actions touch (14).

$$L.(v,u) \rightarrow J.(u,v). \tag{15}$$

This property holds initially and is apparently preserved by occurrence of $e(u,v)$. $L.(v,u)$ prevents $f(u,v)$ by (10). No other occurrences of actions touch (15).

79.3 Priority among neighbors

In each reachable state, neighboring sites are related by *priority*: A site u has priority over its neighbor v iff v has been *critical* more recently. Hence, u gains priority over v upon occurrence of $c(v)$, and looses priority upon occurrence of $c(u)$. No other action affects priority among neighbors u and v. Consequently, the effect of $c(v)$, which is $D.v$, immediately shows priority of u over v. Occurrence of $d(v)$ does not effect priority, hence $F.(v,u)$ also shows priority of u over v. Likewise, occurrences of $f(v,u)$ and $g(v,u)$ retain

u's priority over v. Their effect is $G.(v,u) \wedge K.(v,u)$ and $G.(v,u) \wedge M.(v,u)$. Both formulas imply $G.(v,u) \wedge \neg J.(v,u)$ (by (8) and (7), respectively), which will turn out sufficient to characterize priority. Finally, $g(v,u)$ retains u's priority over v, yielding $G.(u,v)$. This action can occur only in the context of $J.(u,v)$. Altogether, priority of some site u over one of its neighbors v is defined by

$$\text{nprior}(u,v) \text{ iff}$$
$$D.v \vee F.(v,u) \vee (G.(v,u) \wedge \neg J.(v,u)) \vee (G.(u,v) \wedge J.(u,v)). \tag{16}$$

In the rest of this section we will prove that nprior in fact is well defined, i.e., exactly one of two neighbors u and v has priority at each reachable state; formally

$$\text{nprior}(u,v) \text{ iff } \neg\text{nprior}(v,u). \tag{17}$$

This is equivalent to the two propositions

$$\text{nprior}(u,v) \rightarrow \neg\text{nprior}(v,u). \tag{18}$$

and

$$\neg\text{nprior}(v,u) \rightarrow \text{nprior}(u,v). \tag{19}$$

The following shorthands will simplify proof of (18) and (19): Let

$\alpha := D.v \vee F.(v,u) \vee (G.(v,u) \wedge \neg J.(v,u))$,
$\beta := G.(u,v) \wedge J.(u,v)$,
$\gamma := D.u \vee F.(u,v) \vee (G.(u,v) \wedge \neg J.(u,v))$,
$\delta := G.(v,u) \wedge J.(v,u)$.

Then, (18) is equivalent to $\neg(\alpha \wedge \gamma) \wedge \neg(\alpha \wedge \delta) \wedge \neg(\beta \wedge \gamma) \wedge \neg(\beta \wedge \delta)$. The first and third sub-formula, $\neg(\alpha \wedge \gamma)$ and $\neg(\beta \wedge \gamma)$, follow from the invariant property (9). The second and third sub-formulas are propositional tautologies.

Correspondingly, (19) is equivalent to $\alpha \vee \beta \vee \gamma \vee \delta$, which in turn follows from (9).

Priority changes only upon occurrence of transition c: Let $r \xrightarrow{t} s$ be a step. Then

$$r \models \text{nprior}(u,v) \text{ implies } s \models \text{nprior}(u,v) \text{ or } t = c(u). \tag{20}$$

Upon proving (20), assume $r \models \text{nprior}(u,v)$. Then (16) implies four cases:

i. $r \models D.v$. Then $t = d(v)$ yields $s \models F.(v,u)$; hence (20).

ii. $r \models F.(v,u)$. Then $t = f(v,u)$ yields $s \models G.(v,u) \wedge K.(v,u)$; hence $s \models G.(v,u) \wedge \neg J.(v,u)$ with (8), swapping u and v; hence (20). $t = h(v,u)$ yields $s \models G.(v,u) \wedge M.(v,u)$; hence $s \models G.(v,u) \wedge \neg J.(v,u)$ with (7), swapping u and v; hence (20).

iii. $r \models G.(v,u) \wedge \neg J.(v,u)$. Then $t = g(v,u)$ yields $s \models G.(u,v)$; furthermore, enabling of $g(v,u)$ requires $r \models L.(v,u)$, hence $r \models J.(u,v)$ by (15), hence $s \models G.(u,v) \wedge J.(u,v)$, hence (20). $t = c(v)$ yields $s \models D.v$, hence (20). $t = e(v,u)$ is prevented by $G.(v,u)$ and (10), swapping u and v.

iv. $r \models G.(u,v) \wedge J.(u,v)$. Then $t = g(u,v)$ is prevented by $J.(u,v)$ and (8). $t = f(u,v)$ is prevented by $G.(u,v)$ and (9).

No other occurrences of actions $t \neq c(u)$ affect $r \models \text{nprior}(u,v)$. This completes proof of (20).

For a step $r \xrightarrow{t} s$ likewise holds

$$s \models \text{nprior}(u,v) \text{ implies } r \models \text{nprior}(u,v) \text{ or } t = c(v), \tag{21}$$

which can be proven in analogy to (20).

Finally, for a step $r \xrightarrow{t} s$ with $r \not\models \text{nprior}(u,v)$ and $s \models \text{nprior}(u,v)$ holds

$$\text{for all } w \in r(v), \ s \not\models \text{nprior}(v,w). \tag{22}$$

The assumption of (22) with (21) imply $t = c(v)$; then (22) follows with (16).

As a technicality, it will turn out convenient to assign each site u the set $\pi(u)$ of all pairs (u,v), where u has priority over a neighbor, v:

$$\pi(u) := \{u\} \times \text{nprior}(u). \tag{23}$$

70.4 Priority in the network

Priority among any two sites $u, v \in U$ is the transitive closure of priority among neighbors:

$$\text{prior} := \text{nprior}^{+}. \tag{24}$$

As in case of nprior we have to show that prior is well defined, i.e., is asymmetrical:

$$\text{prior}(u,v) \to \neg \text{prior}(v,u). \tag{25}$$

Proof of (25) starts with the initial state $s_{\Sigma_{79.1}}$. For this state, (16) and Fig. 79.1 imply $\text{nprior}(u,v)$ iff $P.(u,v)$, hence $\text{prior}(u,v)$ iff $P^{+}.(u,v)$. Then $P^{+}.(u,v) \to \neg P^{+}.(v,u)$ by the assumption of $x P^{+} y \to x \neq y$, as stated in Fig. 79.1.

Inductively, let $r \xrightarrow{t} s$ be a step and assume $r \models$ (25). If $s \not\models \text{prior}(u,v)$, the proposition $s \models$ (25) is trivial. Otherwise, there exist sites $u_0, \ldots, u_n$, $n \geq 1$ with $u_0 = u$, $u_n = v$, such that with $\beta := u_0 \, \text{nprior} \, u_1 \ldots u_{n-1} \, \text{nprior} \, u_n$ holds: $s \models \beta$. If $r \models \beta$, then $u_0 \neq u_n$ follows from the inductive assumption. Otherwise, there exists an index, i, with $r \not\models u_{i-1} \, \text{nprior} \, u_i$. Then for all $w \in r(u_i)$, $s \not\models \text{nprior}(u_i, w)$, by (22). Then $i = n$, by construction of β. Then there is no $w \in r(v)$ with $s \models v \, \text{nprior} \, w$ (because $u_n = v$). Then $s \models \neg \text{prior}(v,u)$, by (24).

The reverse of (25) is not necessarily valid, i.e., detached sites may be unrelated by priority.

Both above properties imply

$$\text{nprior}(v,u) \to \overline{\text{prior}}(v) \subseteq \overline{\text{prior}}(u), \tag{26}$$

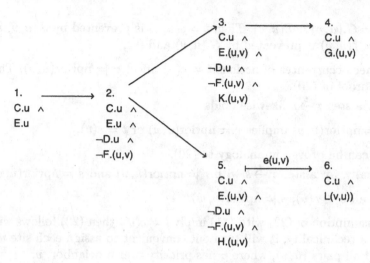

Figure 79.2. Proof graph for (27)

as follows: $\overline{\text{prior}} = \overline{\text{nprior}}^+$ by (24). Then $\overline{\text{prior}}(u) = \bigcup_{v \in \overline{\text{nprior}}(u)} \{v\} \cup$ $\overline{\text{nprior}}(v)$. Furthermore, $\overline{\text{prior}}(u,v) \rightarrow \neg\overline{\text{prior}}(v,u)$, by (25), and $v \in \overline{\text{nprior}}(u)$ iff nprior(v,u). This implies (26).

79.5   Demanded resources

Issuing a *demand* for its resources is the first step of a site on its way to *critical*. A *demanded* resource (u,v) of a *pending* site u will eventually be available to u, or u will send a *message* (v,u) to v:

$$C.u \wedge E.(u,v) \mapsto C.u \wedge (L.(v,u) \vee G.(u,v)). \tag{27}$$

The proof graph of Fig. 79.2 proves (27). Its nodes are justified as follows:

node 1: by (4),
node 2: by (6),
node 3: by (14),
node 5: $H.(u,v)$ excludes $c(u)$ by (10), $\neg D.u$ excludes $d(u)$, $\neg F.(u,v)$ excludes $h(u,v)$.

The formula (27) is now embedded into a context, α. The formula α addresses the set $\pi(u)$ of resources of u for which u has priority. Some of them are *available* to u for its *first time* after being used by v. They constitute a distinguished set, Q. Furthermore, a priorized resource (u,v) is assumed. So, let

$$\alpha := \pi(u) = R \wedge Q \cup \{(u,v)\} \subseteq \pi(u) \wedge G.Q \wedge J.Q. \tag{28}$$

Priority for u may increase whenever one of its neighbors goes critical:

$$\alpha \xrightarrow{c(w)} \pi(u) = R \cup \{(u,w)\} \text{ for all } w \in r(u), \tag{29}$$

which follows directly from (16) and the structure of $\Sigma_{79.1}$.

A step $r \xrightarrow{t} s$ affects α only if u or one of its neighbors goes critical:

$$\alpha \xrightarrow{t} \alpha \text{ for all } t \neq c(w), \text{ with } w \in r(u) \cup \{u\}. \tag{30}$$

(20) and (21) imply that $\pi(u)$ is not touched by occurrence of t. Furthermore, $G.Q$ prevents $f(u,v)$ by (9), and $J.Q$ prevents $g(u,v)$ by (8) for all $v \in r(u)$; hence (30).

Context α yields a further alternative result for (27), with u gaining priority over more neighbors:

$$C.u \wedge \alpha \wedge E.(u,v) \mapsto (C.u \wedge \alpha \wedge (L.(v,u) \vee G.(u,v))) \vee \pi(u) \supset R. \tag{31}$$

The proof graph of Fig. 79.2 can systematically be turned into a proof graph for (31): Replace each node, n, by $n \wedge \alpha$ and augment the following additional outgoing arcs:

$$n \xrightarrow{c(w)} 9.D.w \rightarrow 10.\pi(u) = R \cup \{(u,w)\} \tag{32}$$

with $w \subset \overline{\text{prior}}(u)$.

Additionally, extend justification of each node, n, as follows: $J.(u,v)$ prevents $g(u,w)$ by (8) (replacing v by w), and $G.(u,v)$ prevents $f(u,w)$ by (10) (replacing v by w). The proposition then follows with (30). Node 9 is justified by (29).

79.6   Messages are eventually considered

A site v holding a resource (u,v) without priority for the resource, wil eventually hand it over to u upon request of u, i.e., upon a *message* (v,u). Consideration of the message is a matter of *fairness* of v.

We start with a technicality, crucial for the forthcoming fairness argument: A *repeatedly* used resource is eventually available:

$$K.(u,v) \mapsto G.(u,v), \tag{33}$$

which holds due to the proof graph

with the following justification of nodes:

    node 1: by (14);
    node 2: by pick-up rule of Sect. 64.1;
    node 3: by (5);
    node 4: by pick-up rule of Sect. 64.2; $E.(u,v)$ prevents $f(u,v)$ by (7);

node 5: by pick-up rule of Sect. 64.2; $J.(u, v)$ prevents $h(u, v)$ by (7).

Each message of a priorized site is eventually granted by the respective neighbor:

$$C.u \wedge \text{prior}(u, v) \wedge L.(v, u) \mapsto C.u \wedge G.(u, v). \tag{34}$$

1. $C.u \wedge \text{prior}(u, v) \wedge L.(v, u) \rightarrow$

2. $C.u \wedge \text{prior}(u, v) \wedge L.(v, u) \wedge D.v \overset{d(v)}{\mapsto}$

3. $C.u \wedge \text{prior}(u, v) \wedge L.(v, u) \wedge F.(v, u) \rightarrow$

4. $C.u \wedge \text{prior}(u, v) \wedge L.(v, u) \wedge F.(v, u) \wedge (E.(v, u) \vee J.(v, u)) \mapsto$

5. $C.u \wedge \text{prior}(u, v) \wedge L.(v, u) \wedge G.(v, u) \rightarrow$

6. $C.u \wedge L.(v, u) \wedge G.(v, u) \wedge \neg J.(v, u) \rightarrow$

7. $C.u \wedge L.(v, u) \wedge K.(v, u) \mapsto$

8. $C.u \wedge G.(u, v)$

Figure 79.3. Proof graph for (34)

The proof graph of Fig. 79.3 proves (34). Its nodes are justified as follows:

node 1: by (16); $L.(v, u) \rightarrow \neg G.(u, v)$, by (10);

node 2: $D.v$ excludes $c(u)$, $g(v, u)$, and $c(v)$, by (9);

node 3: $D.v \rightarrow \neg M.(v, u)$ by (4) (swapping u and v), hence the proposition by (7) (swapping u and v);

node 4: by occurrence of $f(v, u)$ or $g(v, u)$; $F.(v, u)$ excludes $c(u)$, $g(v, u)$, and $c(v)$, by (9);

node 5: $G.(v, u)$ implies $\neg D.v \wedge \neg F.(v, u) \wedge \neg G.(u, v)$, by (9), hence the proposition by (16);

node 6: $L.(v, u) \wedge \neg J.(v, u)$ imply $K.(v, u)$, by (11);

node 7: Fairness rule of Theorem 66.3 and (33).

In analogy to the step from (27) to (31), formula (34) can be embedded into the context α. Again, gaining priority over more neighbors arises as an additional alternative:

$$C.u \wedge \alpha \wedge L.(v, u) \mapsto (C.u \wedge \alpha \wedge G.(u, v)) \vee \pi(u) \supset R. \tag{35}$$

The proof graph of Fig. 79.3 can systematically be turned into a proof graph for (35) in exact correspondence to (32), including the extended justification of nodes as given for (32). Fairness rule of Theorem 66.3 is then to be replaced by Corollary 66.4.

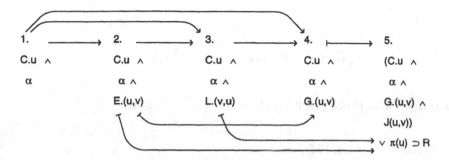

Figure 79.4. Proof graph for (36)

79.7   A pending site obtains its priorized resources

A resource with priority for u is eventually available for u, retaining all already available resources. Alternatively, u may gain priority over more resources:

$$C.u \wedge \alpha \mapsto (C.u \wedge \alpha \wedge G.(u,v) \wedge J.(u,v)) \vee \pi(u) \supset R. \tag{36}$$

Figure 79.4 provides a proof graph for (36). Its nodes are justified as follows:

node 1: by (13);
node 2: by (31);
node 3: by (35);
node 4: by (16), (5), (9).

Iteration of (36) may extend $\pi(u)$, but this is limited by $r(u)$. As α retains all resources in $G \cap J$, all resources with priority for u are eventually available to u:

$$C.u \mapsto C.u \wedge G.\pi(u) \wedge J.\pi(u). \tag{37}$$

The following proof graph proves (37):

$$1.\ C.u \to 2.\ C.u \wedge \alpha \mapsto 3.\ C.u \wedge G.\pi(u) \wedge J.\pi(u). \tag{38}$$

Its nodes are justified as follows:

node 1: Def. (28);
node 2: at most $|r(u)|$ iteration of (36).

79.8   A pending site goes critical

Each *pending* site u may lack priority over some neighbors. It nevertheless goes eventually *critical*: u either gains priority over all its resources, or goes *critical* with some resources over which u has no priority:

$$C.u \mapsto D.u. \tag{39}$$

This will be proven by induction on $|\overline{\text{prior}}(u)|$. The proof graph of Fig. 79.5 covers the case of $\overline{\text{prior}}(u) = \emptyset$. Its nodes are justified as follows:

$$1. \quad \rightarrow \quad 2. \quad \mapsto \quad 3. \quad \overset{c(u)}{\mapsto} 4.$$

$$C.u\wedge \qquad C.u\wedge \qquad C.u\wedge \quad D.u$$

$$\overline{\mathrm{prior}}(u) = \emptyset \quad \pi(u) = r(u) \quad G.r(u)\wedge$$
$$J.r(u)$$

Figure 79.5. Proof graph for the induction basis of (39)

node 1: by (17), (23);
node 2: by (37);
node 3: $J.r(u)$ prevents $g(u, v)$ by (8), $G.r(u)$ prevents $f(u, v)$ by (6).

Now let $\overline{\mathrm{prior}}(u) = M$ and inductively assume $C.v \mapsto D.v$ whenever $\overline{\mathrm{prior}}(v) \subset M$. Then Fig. 79.6 provides a proof graph for the inductive step. Its nodes are justified as follows:

node 1: by (37);
node 2: propositional logic;
node 3: by pick-up rule of Sect. 64.2; for each $v \in \mathrm{nprior}(u)$, $J.\pi(u)$ prevents $g(u, v)$;
node 4: by (4);
node 5: by (16);
node 6: by (12);
node 7: by (26);
node 8: inductive assumption;

1. $C.u$

    $\downarrow$

2. $C.u \wedge G.\pi(u) \wedge J.\pi(u)$

3. $C.u \wedge G.r(u) \wedge J.\pi(u) \quad\vdash\!\!\!\!\!\!\!\!\qquad\qquad c(u)$

    $\downarrow g(u, v)$

4. $C.u \wedge \mathrm{nprior}(v, u) \wedge \neg G.(u, v)$

5. $C.u \wedge \mathrm{nprior}(v, u) \wedge \neg G.(u, v) \wedge \neg D.u \wedge \neg F.(u, v) \rightarrow$

6. $C.u \wedge \mathrm{nprior}(v, u) \wedge G.(v, u) \wedge J.(v, u) \rightarrow$

7. $C.u \wedge \mathrm{nprior}(v, u) \wedge C.v \rightarrow$

8. $C.u \wedge \overline{\mathrm{prior}}(v) \subset \overline{\mathrm{prior}}(u) \wedge C.v \mapsto$

9. $C.u \wedge D.v \mapsto$

10. $D.u \quad\leftarrow$

Figure 79.6. Proof graph for the induction step of (39)

node 9: at node 8, let $M := \overline{\text{prior}}(u)$. Then at node 9, $\overline{\text{prior}}(u) \subset M$, by (29) and (30). Hence the proposition by inductive assumption.

79.9 Proof of the essential properties

We are now prepared to prove the essential properties (1) and (2). The safety property (1) follows from a place invariant, by means of the following proof graph (in fact, a sequence of implications), with $u \in U$ and $v \in r(u)$:

1) $D.u \rightarrow$ 2) $r(D).(u,v) \rightarrow$ 3) $\neg r(D).(v,u) \rightarrow$ 4) $\neg D.v$.

Justification of nodes:

node 1: by definition of $r(D)$ in Fig. 79.1
node 2: by (9)
node 3: by definition of $r(D)$ in Fig. 79.1

The liveness property (2) is shown by the following proof graph, with $u \in U$:

1) $B.u \mapsto$ 2) $C.u \mapsto$ 3) $D.u$.

Justification of nodes:

node 1: pick-up pattern of Sect. 64.1
node 2: by (39).

80 Consensus in Networks

The essential property of the consensus algorithm of Sect. 35 is stability of consensus:

In case all sites are agreed, no request remained initiated.   (1)

Furthermore, no action is enabled in this case. We discuss this property for all three algorithms of Sect. 35.

80.1 Stability of the basic consensus algorithm

Figure 80.1 recalls the basic consensus algorithm of Fig. 35.1 with renamed places. We have to show:

$$B.U \rightarrow D = \emptyset. \tag{2}$$

To this end we consider two place invariants

$$A + B = U \tag{3}$$

and

$$C + \overline{D} = M, \tag{4}$$

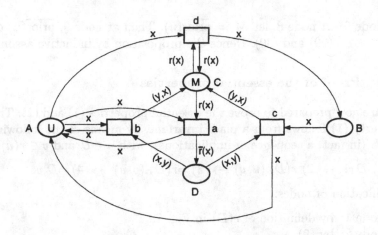

sort	site		fct	r, r̄ :site → set of messages
sort	message = site × site		var	x, y : site
const	U : set of sites			r(x) = {x} × M(x)
const	M : set of messages			r̄(x) = M(x) × {x}

Figure 80.1. Renamed basic consensus algorithm $\Sigma_{35.1}$

as well as the trap

$$r(A) + C \geq r(U). \tag{5}$$

The inscriptions of Fig. 35.1 furthermore imply

$$r(U) = M. \tag{6}$$

Then (3) and (6) imply

$$r(A) + r(B) = r(U) = M. \tag{7}$$

Subtraction of (5) from the sum of (4) and (7) yields (4)+(7)−(5):

$$r(B) + \overline{D} \leq M. \tag{8}$$

Now we conclude

$$B.U \rightarrow r(B).M \rightarrow \overline{D} = \emptyset \tag{9}$$

by (6) and (8). Obviously, $\overline{D} = \emptyset \rightarrow D = \emptyset$, hence (2).

80.2   Stability of the advanced consensus algorithm

Figure 80.2 recalls the advanced consensus algorithm of Fig. 35.2, with renamed places. We are to show three properties.

Firstly, a site u may be *agreed* as well as *demanded* only if an *initiated* message for u is pending, i.e., u not finally *agreed*. In Fig. 80.2 this reads

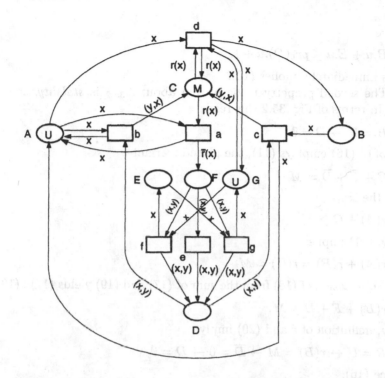

sort	site		fct	r, r̄ :site → set of messages
sort	message = site × site		var	x, y : site
const	U : set of sites			r(x) = {x} × M(x)
const	M : set of messages			r̄(x) = M(x) × {x}

Figure 80.2. Renamed advanced consensus algorithm $\Sigma_{35.2}$

$$B.u \wedge E.u \rightarrow pr_1(D).u. \tag{10}$$

Proof of (10) is again based on two place invariants,

$$A + B = U \tag{11}$$

and

$$E + G = U, \tag{12}$$

as well as the trap

$$A + G + pr_1(D) \geq U. \tag{13}$$

Now, (13) is subtracted from the sum of (11) and (12), yielding (11)+(12) −(13):

$$B + E - pr_1(D) \leq U, \tag{14}$$

hence

$$B.u + E.u \leq pr_1(D).u + 1. \tag{15}$$

This immediately implies (10).

The second property to be proven about $\Sigma_{35.2}$ is *stability*, as stated in (1). In terms of Fig. 35.2 this reads

$$B = U \rightarrow D = \emptyset \tag{16}$$

Proof of (16) employs (11), the place invariant

$$C + \overline{F} + \overline{D} = M \tag{17}$$

and the trap

$$r(A) + C \geq M. \tag{18}$$

Now, (11) implies

$$r(A) + r(B) = r(U) = M \tag{19}$$

and subtraction of (18) from the sum of (17) and (19) yields (17)+(19)−(18):

$$r(B) + \overline{F} + \overline{D} \leq M. \tag{20}$$

Now, definition of r and (20) imply

$$B = U \rightarrow r(B) = M \rightarrow \overline{D} = 0 \rightarrow D = \emptyset, \tag{21}$$

hence (16).

The third property to be proven about $\Sigma_{35.2}$ states that no site is *demanded* (hence each site is *quiet*, by (12)) in case all neighbors are agreed. In terms of $\Sigma_{80.2}$ this reads

$$B = U \rightarrow E = \emptyset. \tag{22}$$

Upon proving (22) we observe that (11), (12), (13), and (16) imply

$$r(A) + r(B) = M, \tag{23}$$

$$r(E) + r(G) = M, \tag{24}$$

$$r(A) + r(G) + D \geq M, \tag{25}$$

$$r(B) = r(U) \rightarrow D = \emptyset. \tag{26}$$

Now, (25) is subtracted from the sum of (23) and (24), yielding (23)+(24)−(25):

$$r(B) + r(E) - D \leq M, \tag{27}$$

hence with (26),

$$r(B) = M \rightarrow r(B) + r(E) \leq M, \tag{28}$$

hence

$$B = U \rightarrow B + E \leq U, \tag{29}$$

which implies (22).

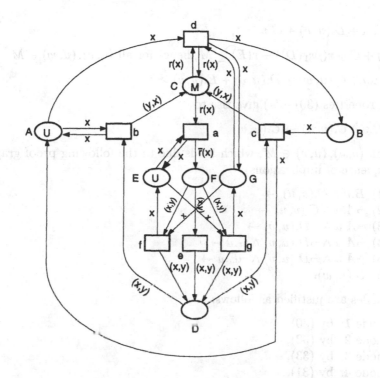

sort	site		fct	r, $\bar{r}$:site $\rightarrow$ set of messages
sort	message = site×site		var	x, y : site
const	U : set of sites			r(x) = {x} × M(x)
const	M : set of messages			$\bar{r}$(x) = M(x) × {x}

Figure 80.3. Renamed variant of the advanced consensus algorithm $\Sigma_{35.3}$

80.3   Stability of the further variant

Figure 80.3 recalls the advanced consensus algorithm, with renamed places.
Its verification is based on three place invariants:
$A + B = U$, hence for each $u \in U$

$$A.u + B.u = 1; \tag{30}$$

$E + G = U$, hence for each $u \in U$

$$E.u + G.u = 1; \tag{31}$$

$\overline{C} + F + D = M$, hence for each $(u, v) \in M$

$$C.(v, u) + F.(u, v) + D.(u, v) = 1; \tag{32}$$

and two traps $A + G + pr_1(D) \geq U$, hence for each $u \in U$ there exists some
$v \in U$ with

$$A.u + D.(u, v) + G.u \geq 1; \tag{33}$$

$r(A) + C + r(pr_1(D)) + r(E) \geq M$, hence for all $(u, v), (u, w) \in M$

$$A.u + C.(u, w) + D.(u, v) + E.(u, v) \geq 1. \tag{34}$$

Properties (30)–(34) give rise to

$$B.u \wedge C.(v, u) \rightarrow C.(u, w) \tag{35}$$

for all $(v, u), (u, v) \in M$, which holds due to the following proof graph (just a sequence of implications):

1) $B.u \wedge C.(v, u) \rightarrow$
2) $\neg A.u \wedge C.(v, u) \rightarrow$
3) $\neg A.u \wedge \neg D.(u, v) \rightarrow$
4) $\neg A.u \wedge \neg D.(u, v) \wedge G.u \rightarrow$
5) $\neg A.u \wedge \neg D.(u, v) \wedge \neg E.u \rightarrow$
6) $C.(u, w)$.

Its nodes are justified as follows:

node 1: by (30),
node 2: by (32),
node 3: by (33),
node 4: by (31),
node 5: by (34).

Now, for $i \in \mathbb{N}$ let

$$M_i := r(\textstyle\bigcup_{j=0}^{i} q^j(u)). \tag{36}$$

We show

$$B.U \wedge C.(v, u) \rightarrow C.M_i \tag{37}$$

by induction on i.
For the induction basis, (35) implies $B.u \wedge C.(v, u) \rightarrow C.r(u)$, which is (37) for $i = 0$. For the induction step, assume (37) and let $(u', w') \in M_{i+1}$. Then there exists some $(v', u') \in M_i$, by construction of M_i. Then

1) $B.U \wedge C.(v, u) \rightarrow$
2) $B.U \wedge C.(v', u') \rightarrow$
3) $C.(u', w')$,

with

node 1: due to the induction hypothesis and
node 2: due to (35).

This goes for each $(u', w') \in M_{i+1}$, hence $B.U \wedge C.(v, u) \to C.M_{i+1}$, which completes the induction step.

Obviously, $M = \bigcup_{i=0}^{\infty} M_i$. Then (37) implies

$$B.U \wedge C.(v, u) \to C.M \tag{38}$$

Furthermore,

$$B.U \wedge C.(v, u) \to C.M \wedge G.U \tag{39}$$

due to the following proof graph (just a sequence of implications):

1) $B.U \wedge C.M \to$
2) $\neg A.U \wedge C.M \to$
3) $\neg A.U \wedge D = \emptyset \to$
4) $G.U$,

with

node 1: by (30).
node 2: by (32).
node 3: by (33).

Stability of $\Sigma_{80.3}$ is now proven by means of an unconventional argument, along interleaved runs:

Let s be a reachable state with $s \models B.U$. Then there exists an interleaved run $s_0 \xrightarrow{t_1} s_1 \xrightarrow{t_2} \ldots \xrightarrow{t_n} s_n$ with $s_n = s$. Then there exists a smallest index, i, with $s_i \models B.U$. Furthermore, $i > 0$, because $\neg(s_0 \models B.U)$. Then $t_i = d(u)$, for some $u \in U$. Then $s_i \models B.U \wedge C.(v, u)$ for each $v \in q(u)$. Then $s_i \models C.M \wedge G.U$, by (39). Then $i = n$, because no action is enabled at s_i. Hence $s \models B.U$ implies $s \models B.U \wedge C.M \wedge G.U$. This with (32) implies

$$B.U \to D = \emptyset \wedge G.U$$

which is the stability property for $\Sigma_{80.3}$.

81 Phase Synchronization on Undirected Trees

81.1 Properties to be proven

Figure 81.1 is a redrawn version of the phase synchronizations algorithm of Fig. 36.1. It has two decisive properties: Firstly, busy sites are always in the same round. In Fig. 81.1 this reads

$$A.(u, n) \wedge A.(v, m) \to n = m \tag{1}$$

and is a typical state property. Secondly, each site will consecutively increase its round number. In Fig. 81.1 this reads

$$A.(u, n) \mapsto A.(u, n + 1) \tag{2}$$

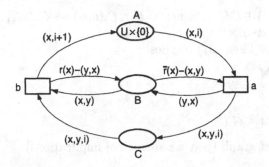

sort   site
sort   message = site × site × nat
const U: set of sites
const W : set of (sites × sites)
fct   r, $\overline{r}$: site × nat → set of messages
var   x, y : site
var   i : nat

$W = W^{-1}$

$x,y \in U \rightarrow x\, W^{*}\, y$

$W_1 = U$

$x_0 W x_1 \ldots x_n W x_{n+1} \wedge$
$x_{i-1} \neq x_{i+1}$ for i=1,...,n
$\rightarrow x_0 \neq x_n$

$r(x) = W(x) \times \{x\}$

$\overline{r}(x) = r(x)^{-1}$

Figure 81.1. Renamed phase synchronization, $\Sigma_{36.1}$

for each $u \in U$ and each $n \in \mathbb{N}$, hence a typical liveness property. Both (1) and (2) will be verified in the sequel.

As a matter of convenience we slightly generalize the conventions of Sect. 31.4 for projection of pairs and triples:

$$(a, b, c)_1 = a$$

and

$$(a, b, c)_{1,2} = (a, c, b)_{1,3} = (b, a, c)_{2,1} = (a, b).$$

Furthermore, four functions $\alpha, \overline{\alpha}, \beta, \overline{\beta} : U \times \mathbb{N} \rightarrow \mathfrak{M}(U \times U)$ will be employed, defined by

$$\alpha(u, n) := 2n \cdot r(u),$$

$$\overline{\alpha}(u, n) := 2n \cdot \overline{r}(u),$$

$$\beta(u, n) := (2n + 1) \cdot r(u),$$

$$\overline{\beta}(u, n) := (2n + 1) \cdot \overline{r}(u)$$

Those four functions, as well as the above generalized projections, will canonically be lifted to multisets, as described in Def. 58.6.

81.2   Place invariants

$\Sigma_{81.1}$ has four important place invariants. Three of them are quite intuitive. First of all, $|A| + |C| = |U|$, which immediately implies

$$|C_1| = |C_{1,2}|, \tag{3}$$

hence each site u has always a unique round number, and if pending, it is pending with a unique site, v.

Secondly, $A_1 + C_1 = U$, which for each $u \in U$ implies

$$A_1.u + C_1.u = 1. \tag{4}$$

Hence each site is always either busy or pending.

The place invariant $B + \overline{B} + r(C_1) + \overline{r}(C_1) = 2(C_{1,2} + C_{2,1})$ relates pending neighbors to their mutual messages. For each pair (u, v) of neighboring sites this implies

$$B.(u, v) + B.(v, u) + r(C_1).(u, v) + r(C_1).(v, u)$$
$$= 2 \cdot C_{1,2}.(u, v) + 2 \cdot C_{1,2}.(v, u). \tag{5}$$

Furthermore, $\neg C_1.u \wedge C_1.v$ implies $r(C_1).(u, v) = C_{1,2}.(u, v) = 0 \wedge r(C_1).(v, u) = C_{1,2}.(v, u) = 1$, hence $B.(u, v) + B.(v, u) = 1$, by (5), hence with (4),

$$A_1.u \wedge \neg A_1.v \rightarrow B.(u, v) \wedge B.(v, u). \tag{6}$$

The place invariant above furthermore implies

$$|B| + |\overline{B}| = 2|C_{1,2} + C_{2,1}| - |r(C_1)| - |\overline{r}(C_1)|. \tag{7}$$

The fourth place invariant is $\alpha(A) + \overline{B} + \beta(C_{1,3}) = \overline{\alpha}(A) + B + \overline{\beta}(C_{1,3})$, which implies for all $u, v \in U$:

$$\alpha(A).(u, v) + B.(v, u) + \beta(C_{1,3}).(u, v) =$$
$$\alpha(A).(v, u) + B.(u, v) + \beta(C_{1,3}).(v, u). \tag{8}$$

This invariant links all places of $\Sigma_{81.1}$.

81.3   Busy neighbors don't exchange messages

In case two neighboring sites u and v are both busy, there is no message available from u to v or from v to u. In terms of $\Sigma_{81.1}$ this reads for all $u \in U$ and $v \in q(u)$:

$$A_1.u \wedge A_1.v \rightarrow B.(u, v) = B.(v, u) = 0. \tag{9}$$

Upon proving (9), assume a state s with $s \models A_1.u \wedge A_1.v$. Then at s holds $A_1.u = A_1.v = 1$, hence $C_1.u = C_1.v = 0$ (by (4)), hence $C_{1,2}.(u, v) = C_{1,2}.(v, u) = 0$ (by (3)), hence the proposition, by (5).   □

81.4   A property of neighboring pending sites

A neighbor v of a pending site u is pending with u, or u is pending with v. In terms of $\Sigma_{81.1}$, for $u \in U$ and $v \in q(u)$,

$$C_1.u \to C_{1,2}.(u, v) \lor C_{1,2}.(v, u). \tag{10}$$

Proof of (10) assumes a state s with $s \models C_1.u = 1$. Then at s holds for all $w \in q(u) : r(C_1).(u, w) = 1$, hence particularly $r(C_1).(u, v) = 1$, hence $C_{1,2}.(u, v) + C_{1,2}.(v, u) \geq 1$, by (5), hence the proposition.   □

81.5   A site is pending with a busy neighbor

A pending site v with a busy neighbor u is pending with u. (Hence, with (3), at most one neighbor of a pending site is busy). In terms of $\Sigma_{81.1}$, for $u \in U$ and $v \in q(u)$,

$$A_1.u \land C_1.v \to C_{1,2}.(v, u). \tag{11}$$

Proof of (11) combines two properties of $\Sigma_{81.1}$: First, $C_1.v$ implies $C_{1,2}.(u, v) \lor C_{1,2}.(v, u)$ by (10). Second, $A_1.u$ implies $\neg C_1.u$ by (4), hence $\neg C_{1,2}.(u, v)$.   □

81.6   Three pending neighbors form a sequence

Assume a site v, pending with w. Then each other pending neighbor u of v is pending with v. In $\Sigma_{81.1}$ this reads for $v \in U$ and $u, w \in q(v)$:

$$C_1.u \land C_{1,2}.(v, w) \to C_{1,2}.(u, v). \tag{12}$$

Proof of (12) combines two properties of $\Sigma_{81.1}$: First, $C_1.u$ implies $C_{1,2}.(u, v) \lor C_{1,2}.(v, u)$ by (10). Second, $C_{1,2}.(v, w)$ implies $\neg C_{1,2}.(v, u)$, by (4).   □

81.7   Busy neighbors are in the same round

If two neighbors u and v are both busy, they operate in the same round. In $\Sigma_{81.1}$ this reads for $u \in U$, $v \in q(u)$, and $n, m \in \mathbb{N}$:

$$A.(u, n) \land A.(v, m) \to n = m. \tag{13}$$

To prove (13), let s be a reachable state of $\Sigma_{81.1}$ with $s \models A_1.u \land A_1.v$. Then at s holds $C_1.u = C_1.v = 0$ by (4), hence $\beta(C_{1,3}).(u, v) = \beta(C_{1,3}).(v, u) = 0$. Furthermore, $B.(u, v) = B.(v, u) = 0$, by (9). Combining both properties, (8) yields $\alpha(A).(u, v) = \alpha(A).(v, u)$. Then for each $n \in \mathbb{N}$, $A.(u, n) = A.(v, n)$. Then (13) follows with (4).   □

81.8   A property of chains

Given $u_0, \ldots, u_n \in U$, the sequence $u_0 \ldots u_n$ is a *chain* if $u_{i-1} \in r(u_i)$ for $i = 1, \ldots, n$, and $u_{i-1} \neq u_{i+1}$ for $i = 1, \ldots, n-1$.

Assume a chain $u_0 \ldots u_n$, starting with a busy site, u_0, followed by a pending site, u_1. Then all follower sites $u_2, \ldots, u_n$ are pending. In $\Sigma_{81.1}$ this reads

$$A_1.u_0 \wedge C_1.u_1 \rightarrow C_1.u_i \text{ for all } i = 1, \ldots, n. \tag{14}$$

To prove (14), let s be a reachable state with $s \models A_1.u_0 \wedge C_1.u_1$. Then at s holds $C_{1,2}.(u_1, u_0)$ by (11). Then

$$\neg C_{1,2}.(u_1, u_2) \tag{*}$$

by (4). Now, contradicting (14), assume an index $1 \leq i \leq n$ with $s \models \neg C_1.u_i$. Let j be the smallest of those indices. Then at s holds $A_1.u_j$ by (4), hence $C_{1,2}.(u_{j-1}, u_j)$, by (11). Then $C_{1,2}.(u_{i-1}, u_i)$ for $i = 2, \ldots, n$ by iterated application of (12). Then in particular $C_{1,2}.(u_1, u_2)$, which contradicts (*).

□

81.9   Proof of the state property (1)

We are now prepared to prove (1) as follows:

Let s be a reachable state with $s \models A.(u, n) \wedge A.(v, m)$. Then there exists a chain $u_0 \ldots u_n$ in U with $u_0 = u$ and $u_n = v$. Then $s \models A_1.u_i$ for $i = 0, \ldots, n$, by (14) and (4). Then at s holds $A.(u_i, n)$ for $i = 0, \ldots, n$ by iteration of (13). Hence $n = m$.

81.10   Pending sites have pending messages

Here we start proof of the liveness property (2). First, we observe pending messages in case all sites are pending:

$$C_1.U \rightarrow |B| > 0. \tag{15}$$

Proof of (15) is based on the observation that an undirected tree with n nodes has $n - 1$ arcs. Hence, in $\Sigma_{81.1}$,

$$|r(U)| = |\bar{r}(U)| = |U| - 1. \tag{16}$$

Then $C_1.U \rightarrow |B| + |\bar{B}| = 2|C_{1,2} + C_{2,1}| - |r(C_1)| - |\bar{r}(C_1)|$ (by (7)) $= 2|U| - 2(|U| - 1)$ (by (16)) $= 2$.

81.11 $\Sigma_{81.1}$ is deadlock free

Each reachable state of $\Sigma_{81.1}$ enables at least one action.   (17)

Proof. Let s be a reachable state of $\Sigma_{81.1}$. 1st case: $s \models A_1.u$ for at least one $u \in U$. Then there exists a chain $u_0 \ldots u_n$, $n \geq 0$, of sites with $s \models A_1.u_i$ for all $i = 0, \ldots, n$, and $\neg A.v$ for all $v \in q(u_n) - u_{n-1}$. Hence for all such v holds $s \models B.(v, u) \vee B.(u, v)$, by (6). Now we distinguish two cases: Firstly, $s \models B.(u, v)$ for all $v \in q(u_n) - u_{n-1}$. Then s enables $a(u_n, u_{n-1}, k)$, where $s \models A.(u_n, k)$. Otherwise, there exists some $v \in r(u_n) - u_{n-1}$ with $s \models B.(v, u)$. Furthermore, $s \models C.(v, u, k)$ for some $k \in \mathbf{N}$ (with (4)). Then s enables $b(v, u, k)$. 2nd case: There is no $u \in U$ with $s \models A_1.u$. Then $s \models C_1.U$ (with (4)). Then $|B| > 0$, by (15). Hence there exist $u, v \in U$ with $s \models B.(u, v)$. Then $s \models C.(u, v, k)$ for some $k \in \mathbf{N}$, by (5). Then s enables $b(u, v, k)$. $\square$

81.12 The weight function γ

A function $\gamma(u, v)$ will be considered, which for neighbors u and v yields an integer value $\gamma(u, v)$ at any given state s. Values $\gamma(u_{i-1}, u_i)$ remain in a limited interval for all chains $u_0 \ldots u_n$, and occurrences of transitions increase those values. For $u, v \in U$, let

$$\gamma(u, v) := B.(v, u) + \Sigma_{n \in \mathbf{N}} 2n \cdot A.(u, n) + (2n + 1) \cdot C_{1,3}.(u, n).   (18)$$

Then (8) implies

$$\gamma(u, v) = \gamma(v, u).   (19)$$

Furthermore, for neighbors w of u, $C_{1,2}.(u, w)$ iff $r(C_1).(u, w)$; hence $B.(w, u) \leq 2$ (by (5)), hence

$$|\gamma(u, v) - \gamma(u, w)| \leq 2,   (20)$$

again by (5). Then for each sequence $u_0 \ldots u_k$ of sites, (19) and (20) imply

$$|\gamma(u_0, u_1) - \gamma(u_{n-1}, u_n)| \leq 2(k - 1).   (21)$$

81.13 Proof of the liveness property (2)

Inspection of $\Sigma_{81.1}$ yields for each step $r \overset{t}{\rightarrow} s$ with $t = a(u, v, i)$ or $t = b(u, v, i)$:

If $\gamma(u, v) = n$ at state r, then $\gamma(u, v) > n$ at state s.   (22)

Property (17) implies at least one pair (u, v) of neighbors with infinitely many occurrences of $a(u, v, i)$ and $b(u, v, i)$. Then in the set of all reachable states, $\gamma(u, v)$ is not limited, by (22). This applies to all neighbors u, v, by (21). Hence (2).

82   Distributed Self-Stabilization

82.1   Properties to be proven

Figure 82.1 is a redrawn version of the distributed self stabilization algorithm. We have to show that the overall workload remains constant, eventually is balanced, and henceforth remains balanced.

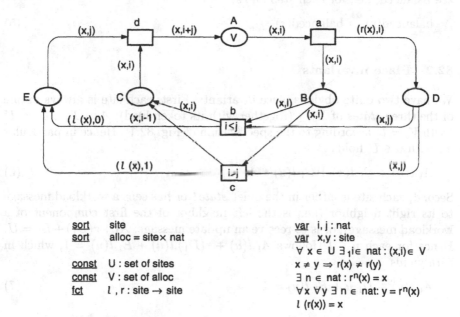

<pre>
sort   site   var i, j : nat
sort   alloc = site× nat   var x,y : site
                                      ∀ x ∈ U ∃₁i∈ nat : (x,i)∈ V
const   U : set of sites   x ≠ y ⇒ r(x) ≠ r(y)
const   V : set of alloc   ∃ n ∈ nat : rⁿ(x) = x
fct   l , r : site → site   ∀x ∀y ∃ n ∈ nat: y = rⁿ(x)
                                      l (r(x)) = x
</pre>

Figure 82.1. Renamed distributed load balancing

A formal representation of those properties in terms of $\Sigma_{82.1}$ can be based on the following functions. For any place $p \in \{A, B, C, E\}$ and any site $u \in U$, let

$$\sigma(p, u) := \begin{cases} 0 & \text{iff } \neg p_1.u \\ n & \text{iff } p.(n, u) \end{cases},$$

$$\sigma(u) := \Sigma_{\{A,B,C,E\}}\sigma(p, u), \text{ and} \qquad (1)$$

$$\sigma := \Sigma_{u\in U}\sigma(u).$$

These functions describe the workload of site u at place p, the entire workload of u and the overall workload in the system, respectively. The initial overall workload is k iff $a_{\Sigma_{82.1}} \models \sigma = k$. A *balanced* state meets the predicate

$$balanced := u, v \in U \to |\sigma(u) - \sigma(v)| \leq 1. \qquad (2)$$

So we have to show the state property

$$\Sigma_{82.1} \models \sigma = k \tag{3}$$

and the progress property

$$\Sigma_{82.1} \models a_\Sigma \mapsto \text{balanced.} \tag{4}$$

Furthermore, we have to show that all states reachable from a balanced state are balanced, i.e., for each step $r \xrightarrow{t} s$,

$$\text{balanced}(r) \rightarrow \text{balanced}(s). \tag{5}$$

82.2   Place invariants

We have two quite obvious place invariants. First, each site is always in one of the three *states* of $\Sigma_{37.1}$ (together with its token load): $A_1 + B_1 + C_1 = U$ (with $V_1 = U$ according to the specification of Fig. 82.1). Hence in particular for each $u \in U$ holds

$$A_1.l(u) + B_1.l(u) + C_1.l(u) = 1. \tag{6}$$

Second, each site is either in the quiet *state1* or has sent a workload message to its right neighbor (i.e., is the left neighbor of the first component of a workload message), or is to receive an update message: $A_1 + r(D_1) + E_1 = U$. Hence for each $u \in U$ follows $A_1.l(u) + r(D_1).l(u) + E_1.l(u) = 1$, which in turn yields

$$A_1.l(u) + D_1.u + E_1.l(u) = 1. \tag{7}$$

82.3   Further properties of $\Sigma_{82.1}$

Two basic properties are required in the sequel: The ground formula $A_1.U$, and an upper bound for the workload of the sender of a workload message. To start with, we first show

$A_1.U$ is a ground formula. \hfill (8)

Upon proving (8), observe that all steps starting at $A_1.U$ are shaped $A_1 \xrightarrow{a(u,n)} A_1.U - u \wedge B_1.u \wedge D_1.r(u)$, for some $u \in U$ and $n \in \mathbb{N}$. Then (8) follows from Theorem 70.2 and the following proof graph:

1) $A_1.U - u \wedge B_1.u \wedge D_1.r(u) \hookrightarrow$
2) $B_1.U \wedge D_1.U \hookrightarrow$
3) $C_1.U \wedge E_1.U \hookrightarrow$
4) $A_1.U.$

Its nodes are justified by the pick-up pattern of Sect. 69.1 together with the following:

1) by occurrence of $a(v, n)$ for all $(v, n) \in V, v \neq u$
2) by occurrence of $b(v, n, m)$ or $c(v, n, m)$ for all $v \in U$
3) by occurrence of $d(v, n, m)$ for all $v \in U$.

Second, we show that a workload message tops its sender's token load:

$$D.(u, n) \rightarrow \sigma(l(u)) \leq n. \tag{9}$$

(9) is obviously true at the initial state. Inductively assume a step $r \xrightarrow{t} s$ with $r \models$ (9). Upon proving $s \models$ (9) two cases are distinguished:

i. Assume $r \not\models D.(u, n)$ and $s \models D.(u, n)$. Then $t = a(l(u), n)$ (by the structure of the net). Then $s \models B.(l(u), n)$ (by the occurrence rule). Hence $s \models B_1.l(u)$, hence $s \models \neg A_1.l(u) \wedge \neg C_1.l(u)$, by (6). Furthermore, the assumption of $s \models D.(u, n)$ implies $s \models D_1.u$, hence $s \models \neg E_1.l(u)$ (by (7)). Both arguments together imply $\sigma(l(u)) \leq \sigma(B.l(u))$. Then $s \models B.(l(u), n)$ implies the proposition.

ii. Assume $r \models \sigma(l(u)) \leq n$ and $s \not\models \sigma(l(u)) \leq n$. Then $t = c(u, n, m)$, for some $n, m \in N$ (by the structure of the net). Then $s \models E_1.(l(u), n)$ (by the occurrence rule). Then $s \models \neg D_1.(u, n)$ (by (7)), hence the proposition.

82.4 A decreasing weight

A weight function τ on states will be employed, defined for each state s of $\Sigma_{82.1}$ by $\tau(s) = n$ iff $s \models$

$$\Sigma_{u \in U} \sigma(u)^2 = n. \tag{10}$$

It will turn out that no step increases τ. Furthermore, τ decreases upon occurrence of $c(u, n, m)$, provided $m + 1$ is smaller than n.

First we show that c_3 does not increase τ: Let $r \xrightarrow{c(u,n,m)} s$ be a step. Then

$$\tau(r) \geq \tau(s). \tag{11}$$

In order to show (11), observe that at r holds [*] $B.(u, n)$ as well as [**] $D.(u, m)$, due to the occurrence rule. Furthermore, with $r \models \sigma(l(u)) = a \wedge \sigma(u) = b$, at r holds $b \geq n$ by [*], $n > m$ by inscription of transition c, and $m \geq \sigma(l(u))$, by [**] and (9); hence [***] $(a - b + 1) \leq 0$. Now,

$$\begin{aligned} \tau(s) &= \tau(r) - a^2 - b^2 + (a + 1)^2 + (b - 1)^2 \text{ (by the structure of } c(u, n, m)) \\ &= \tau(r) - a^2 - b^2 + a^2 + 2a + 1 + b^2 - 2b + 1 \\ &= \tau(r) + 2(a - b + 1) \\ &\leq \tau, \text{ by } [***], \text{ hence (11)}. \end{aligned}$$

(11) can be strengthened in case $\sigma(u) > \sigma(l(u)) + 1$: Let $r \xrightarrow{c(u,n,m)} s$ be a step of $\Sigma_{82.1}$ with $m + 1 < n$. Then

$$\tau(r) > \tau(s) \tag{12}$$

Proof of (12) is a slight variant of the above proof graph of (11): $m + 1 < n$ now implies $b > n$, hence $(a - b + 1) < 0$. Then the last two lines read $\tau(r) + 2(a - b + 1) < \tau(r)$.

Generalizing (11), no step at all increases τ: Let $r \xrightarrow{t} s$ be a step of $\Sigma_{82.1}$. Then

$$\tau(r) \geq \tau(s). \tag{13}$$

To prove (13), observe that $\tau(r) \neq \tau(s)$ implies $t = c(u, n, m)$ for some $u \in U$ and $n, m \in \mathbb{N}$, by definition of τ and σ, and the structure of $\Sigma_{82.1}$. Then (13) follows from (11).

82.5 Descents

A descent of length k consists of a sequence $u, l(u), l^2(u), \ldots, l^{k+1}(u)$ of sites, with token loads decreasing by 1 from u to $l(u)$ and by any number from $l^k(u)$ to $l^{k+1}(u)$, and identical token load of $l(u), \ldots, l^k(u)$. More precisely, for any site $u \in U$ and any state s, the *descent of u at s* amounts to k (written: $\delta(u) = k$) iff there exists some $n \in \mathbb{N}$ with

$$\sigma(u) = n + 1, \ \sigma(l^i(u)) = n \quad (i = 1, \ldots, k), \ \sigma(l^{n+1}(u)) \leq n - 1. \tag{14}$$

Figure 82.2 outlines examples.

In general, there may exist states s with undefined descent $\delta(u)$. Even more, obviously holds for all states s of $\Sigma_{82.1}$:

$$s \text{ is balanced iff no site has a descent at } s. \tag{15}$$

In the sequel we will show that large descents reduce to small ones and small descents reduce the weight τ. Each large descent reduces to a smaller one, as exemplified in Fig. 82.2.

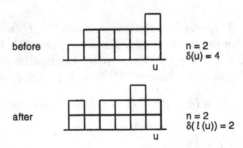

Figure 82.2. Reduction of a large descent

$$A_1.U \wedge \delta(u) = k \wedge k \geq 2 \hookrightarrow A_1.U \wedge \delta(l(u)) = k - 2. \tag{16}$$

This proposition follows from the following proof graph:

1) $A_1.U \wedge \delta(u) = k \wedge k \geq 2 \rightarrow$
2) $A_1.U \wedge A.(u, n+1) \wedge A.(l^i(u), n) \ (i = 1, \ldots, k) \ \wedge A.(l^{k+1}(u), n-j) \hookrightarrow$
3) $B_1.U \wedge D_1.U \wedge B.(u, n+1) \wedge B.(l^i(u), n) \ (i = 1, \ldots, k) \ \wedge D.(u, n) \wedge$
   $D.(l^i(u), n) \ (i = 1, \ldots, k-1) \ \wedge D.(l^k(u), n-j) \hookrightarrow$
4) $C_1.U \wedge E_1.U \wedge C.(l^i(u), n) \ (i = 1, \ldots, k-1) \ \wedge C.(l^k(u), n-j) \wedge$
   $E.(l(u), 1) \wedge E.(l^i(u), 0) \ (i = 2, \ldots, k) \ \hookrightarrow$
5) $A_1.U \wedge A.(l(u), n+1) \wedge A.(l^i(u), n) \ (i = 2, \ldots, k-1) \wedge A.(l^k(u), n-1) \rightarrow$
6) $A_1.U \wedge \delta(l(u)) = k - 2.$

Its nodes are justified as follows:

node 1: there exist $n, j \geq 1$ with the described properties, according to
(14)
node 2: by occurrence of $\{a(v, m) \mid v \in U \wedge A.(v, m)\}$
node 3: by occurrence of $c(u, n+1, n)$, $b(l^i(u), n, n)$ for $i = 1, \ldots, k - 1$, $c(l^k(u), n, n - j)$, and $b(v, m, m')$ or $c(v, m, m')$ for all $v \neq l^i(u) \ (i = 0, \ldots, k)$
node 4: by occurrence of $\{d(v, m, m') \mid v \in U \wedge C.(v, m) \wedge E.(v, m')\}$
node 5: by (14).

Each descent of length 0 reduces the weight τ, as outlined in Fig. 82.3.

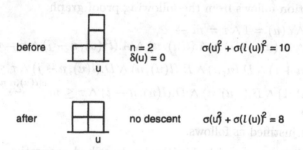

before $n = 2$ $\delta(u) = 0$ $\sigma(u)^2 + \sigma(l(u))^2 = 10$

after   no descent $\sigma(u)^2 + \sigma(l(u))^2 = 8$

Figure 82.3. Descent of length 0

Formally,

$$A_1.U \wedge \delta(u) = 0 \wedge \tau = m \hookrightarrow \tau < m. \tag{17}$$

This proposition follows from the following proof graph:

1) $A_1.U \wedge \delta(u) = 0 \wedge \tau = m \rightarrow$
2) $A.(u, n+1) \wedge A.(l(u), n - j) \wedge \tau = m \overset{c(u,n+1,n-j)}{\hookrightarrow}$
3) $B.(u, n+1) \wedge D.(u, n - j) \wedge \tau \leq m$
4) $\tau < m.$

Its nodes are justified as follows:

node 1: there exist $n, j \geq 1$ with the described properties, according to
(14)

node 2: by occurrence of $a(u, n + 1)$ and $a(l(u), n - j)$
node 3: by (12).

Each descent of length 1 likewise reduces the weight τ, as outlined in Fig. 82.4.

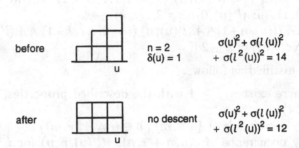

before $n = 2$ $\sigma(u)^2 + \sigma(l(u))^2$
              $\delta(u) = 1$ $+ \sigma(l^2(u))^2 = 14$
u

after   no descent $\sigma(u)^2 + \sigma(l(u))^2$
                            $+ \sigma(l^2(u))^2 = 12$
u

Figure 82.4. Descent of length 1

Formally,

$$A_1.U \wedge \delta(u) = 1 \wedge \tau = m \hookrightarrow \tau < m. \tag{18}$$

This proposition follows from the following proof graph:

1) $A_1.U \wedge \delta(u) = 1 \wedge \tau = m \rightarrow$
2) $A_1.U \wedge A.(u, n + 1) \wedge A.(l(u), n) \wedge A.(l^2(u), n - j) \wedge \tau = m \hookrightarrow$
3) $B.(u, n + 1) \wedge D.(u, n) \wedge B.(l(u), n) \wedge D.(l(u), n - j) \wedge \tau \leq m \overset{c(u, n+1, n)}{\hookrightarrow}$
4) $E.(l(u), 1) \wedge B.(l(u), n) \wedge D.(l(u), n - j) \wedge \tau \leq m \overset{c(l(u), n, n-j)}{\hookrightarrow}$
5) $\tau < m.$

Its nodes are justified as follows:

node 1: there exist $n, j \geq 1$ with the described properties, according to (14)
node 2: by occurrence of $a(u, n + 1)$, $a(l(u), n)$, and $a(l^2(u), n - 1)$
node 3: by the occurrence rule
node 4: by (12).

The weight τ is reducible as long as there exists a descent:

$$\tau = m \hookrightarrow \tau < m \vee \forall u \in U : \delta(u) \text{ is undefined.} \tag{19}$$

This proposition follows from the following proof graph:
1) $\tau = m \hookrightarrow$
2) $A_1.U \wedge \tau \leq m \rightarrow$ 3) $\forall u \in U : \delta(u)$ undefined ┐
   ↓
4) $A_1.U \wedge \tau \leq m \wedge \exists u \in U, k \in \mathbb{N}$ with $\delta(u) = k \hookrightarrow$
5) $A_1.U \wedge \tau \leq m \wedge \exists u \in U$ with $\delta(u) \leq 1 \hookrightarrow$
6) $\tau < m \rightarrow$
7) $\tau < m \vee \forall u \in U : \delta(u)$ undefined ←

Its nodes are justified as follows:

node 1: by (8) and (13)
node 2: propositional logic
node 3: propositional logic
node 4: by $\lfloor \frac{k}{2} \rfloor$ fold application of (16)
node 5: by (17) if $\delta(u) = 0$, and by (18) if $\delta(u) = 1$
node 6: propositional logic.

82.6 Proof of the essential properties

To show (3), let $r \xrightarrow{t} s$ be any step of $\Sigma_{82.1}$, and assume $\sigma_r = k$. Then $\sigma_s = k$ follows due to the structure of $\Sigma_{81.1}$. Finally, (3) follows by induction on the length of interleaved runs of $\Sigma_{82.1}$

To prove (5), first consider the case of $t = c(u, n, m)$ for some $u \in U$ and $n, m \in N$. Then at r holds $B.(u, n) \wedge D.(u, m) \wedge n > m$. Furthermore, $\sigma(u) \geq n$ by (1) and $m \geq \sigma(l(u))$, by (9). Hence $\sigma(u) = n$ and $\sigma(l(u)) = n - 1$, as r is balanced. Then at s holds $\sigma(u) = n - 1$ and $\sigma(l(u)) = n$. The workload $\sigma(v)$ remains unchanged for all $v \neq u$. Hence s is balanced, too.

All actions t not involving c do not touch $\sigma(u)$ for any $u \in U$, hence the proposition.

Proof of (4) requires

$$a_\Sigma \hookrightarrow \text{balanced},\tag{20}$$

proven by the following proof graph:

$$a_\Sigma \rightarrow \tau = m \hookrightarrow \tau = n_1 < m \hookrightarrow \tau = n_2 < n_1 \hookrightarrow \dots \hookrightarrow \tau = n_m = 0$$

$$\searrow \qquad \downarrow \qquad \downarrow \qquad \swarrow$$

$$\forall u \in U : \delta(u) \text{ is undefined}$$

$$\downarrow$$

$$\text{balanced}$$

which is justified as follows: The first implication states that τ has some value, m, at the initial state a_Σ. All other nodes in the upper line are justified by (19). The last implication holds by (15).

In order to show (4), let w be an interleaved run of $\Sigma_{82.1}$. Then there exists a concurrent run K of $\Sigma_{82.2}$, including all actions of w. K has a reachable, balanced state, s, (by (20)). Then w has a reachable state, s', such that all actions of K, occurring before s, are actions of w, occurring before s'. Then $\Sigma_{82.2} \models s \mapsto s'$ and s' is balanced by (5), hence the proposition.

References

[Agh86]   G. A. Agha. *A Model of Concurrent Computation in Distributed Systems*. MIT Press, Cambridge, Mass., 1986.

[AS85]   B. Alpern and F. B. Schneider. Defining liveness. *Information Processing Letters*, 21:181–185, 1985. Safety/Liveness Decomposition.

[BA90]   M. Ben-Ari. *Principles of Concurrent and Distributed Programming*. International Series in Computer Science. Prentice Hall, Englewood Cliffs, N. J., 1990.

[Bar96]   V. Barbosa. *An Introduction to Distributed Algorithms*. MIT Press, Cambridge, Mass., 1996.

[BB82]   G. Berry and G. Boudol. The chemical abstract machine. *TCS*, 1982.

[BCM88]   J.-P. Banâtre, A. Coutant, and D. le Metaye. A parallel machine for multiset transformation and its programming style. *Future Generations Computer Systems*, 4:133–144, 1988.

[BE96]   G. Burns and J. Esparza. Trapping mutual exclusion in the box calculus. *Theoretical Computer Science. Special Volume on Petri Nets*, 153(1–2), January 1996.

[Ben73]   C. H. Bennett. Logical reversibility of computation. *IBM Journal of Research and Development*, 6:525–532, 1973.

[Bes96]   E. Best. *Semantics of Sequential and Parallel Programs*. International Series in Computer Science. Prentice Hall, Englewood Cliffs, N. J., 1996.

[BF88]   E. Best and C. Fernández. *Nonsequential Processes*, volume 13 of *EATCS Monographs on Theoretical Computer Science*. Springer-Verlag, Berlin, 1988.

[BGW89]   G. M. Brown, M. G. Gouda, and C. Wu. Token systems that self-stabilize. *IEEE Transaction on Computers*, 38(6):845–852, 1989.

[BP89]   J. E. Burnes and J. Pachl. Uniform self-stabilizing rings. *ACM Transactions on Programming Languages and Systems*, 11(2):330–344, April 1989.

[Bro87]   M. Broy. Semantics of finite and infinite networks of concurrent communicating agents. *Distributed Computing*, 2:13–31, 1987.

[Cha82]   E. J. H. Chang. Echo algorithms: Depth parallel operations on general graphs. *IEEE Transactions on Software Engineering*, SE-8(4):391–401, 1982.

[CM84]   K. M. Chandy and J. Misra. The drinking philosophers problem. *ACM Transactions on Programming Languages and Systems*, 6(4):632–646, October 1984.

[CM88]   K. M. Chandy and J. Misra. *Parallel Program Design: A Foundation*. Addison-Wesley, Reading, Mass., 1988.

[Des97]   J. Desel. How distributed algorithms play the token game. In C. Freksa, M. Jantzen, and R. Valk, editors, *Foundations of Computer*

Science, volume 1337 of *LNCS Lecture Notes in Computer Science*, pages 297–306. Springer-Verlag, 1997.

[Dij71]   E. W. Dijkstra. Hierarchical ordering of sequential processes. *Acta Informatica*, 1:115–138, 1971.

[Dij74]   E. W. Dijkstra. Self-stabilizing systems in spite of distributed control. *Communications of the ACM*, 17(11):643–644, 1974.

[Dij75]   E. W. Dijkstra. Guarded commands, nondeterminancy, and formal derivation of programs. *Communications of the ACM*, 18(8):453–457, 1975.

[Dij78]   E. W. Dijkstra. Finding the correctness proof of a concurrent program. *Proc. Koninklijke Nederlandse Akademie van Wetenschappen*, 81(2):207–215, 1978.

[DK98]   J. Desel and E. Kindler. Proving correctness of distributed algorithms using high-level Petri nets – a case study. In *1998 International Conference on Application of Concurrency to System Design*, pages 177–186, Fukushima, Japan, March 1998. IEEE Computer Society Press.

[DKVW95]   J. Desel, E. Kindler, T. Vesper, and R. Walter. A simplified proof for a self-stabilizing protocol: A game of cards. *Information Processing Letters*, 54:327–328, 1995.

[DKW94]   J. Desel, E. Kindler, and R. Walter. A game of tokens: A proof contest. *Petri Net Newsletter*, 47:3 – 4, October 1994.

[DS80]   E. W. Dijkstra and C. S. Scholten. Termination detection for diffusing computations. *Information Processing Letters*, 4:1–4, 1980.

[Fin79]   S. G. Finn. Resynch procedures and a fail safe network protocol. *IEEE Transactions on Communications*, COM-27:840–845, 1979.

[FT82]   E. Fredkin and T. Toffoli. Conservative logic. *International Journal of Theoretical Physics*, 21(3/4):219–253, 1982.

[Gan80]   R. Gandy. Church's thesis and principles for mechanisms. In *The Kleene Symposium*, pages 123–274, North-Holland, Amsterdam, 1980. J. Barwise et al., editors.

[GPR97]   J. E. Gehrke, C. G. Plaxton, and R. Rajaraman. Rapid convergence of a local load balancing algorithm for asynchronous rings. In M. Mavronicolas and P. Tsigas, editors, *Distributed Algorithms, WDAG*, volume 1320 of *LNCS Lecture Notes in Computer Science*, pages 81–95. Springer-Verlag, September 1997.

[Har87]   D. Harel. Statecharts: A visual formalism for computer systems. *Science of Computer Programming*, 8(3):231–274, 1987.

[Jen92]   K. Jensen. *Coloured Petri Nets*, volume 1 of *EATCS Monographs on Theoretical Computer Science*. Springer-Verlag, 1992.

[Kin95]   E. Kindler. *Modularer Entwurf verteilter Systeme mit Petrinetzen*. PhD thesis, Technische Universität München, 1995.

[KRVW97]   E. Kindler, W. Reisig, H. Völzer, and R. Walter. Petri net based verification of distributed algorithms: An example. *Formal Aspects of Computing*, 1997.

[KW95]   E. Kindler and R. Walter. Message passing mutex. In J. Desel, editor, *Structures in Concurrency Theory*, Workshops in Computing, pages 205–219, Berlin, May 1995. Springer-Verlag.

[Lam86]   L. Lamport. The mutual exclusion problem: Part I – a theory of interprocess communication. *Journal of the ACM*, 33(2):313–326, 1986.

[Lyn96]   N. A. Lynch. *Distributed Algorithms*. Morgan Kaufmann Publishers, San Francisco, Calif., 1996.

[Mat89]   F. Mattern. *Verteilte Basisalgorithmen*. Informatik-Fachberichte 226, Springer-Verlag, Berlin, 1989.

[Mil89]   R. Milner. *Communication and Concurrency.* International Series in
            Computer Science. Prentice Hall, Englewood Cliffs, N. J., 1989.

[Mis91]   J. Misra. Phase synchronization. *Information Processing Letters,*
            38:101–105, 1991.

[MP92]   Z. Manna and A. Pnueli. *The Temporal Logic of Reactive and Con-
            current Systems.* Springer-Verlag, Berlin, 1992.

[MP95]   Z. Manna and A. Pnueli. *Temporal Verification of Reactive Systems.*
            Springer-Verlag, Berlin, 1995.

[NTA96]   M. Naimi, M. Trehel, and A. Arnold. A log(n) distributed mutual
            exclusion algorithm based on path reversal. *Journal of Parallel and
            Distributed Computing,* 34:1–13, 1996.

[OL82]   S. Owicki and L. Lamport. Proving liveness properties of concur-
            rent programs. *ACM Transactions on Programming Languages and
            Systems,* 4(3):455–495, 1982.

[Pet81]   G. L. Peterson. Myths about the mutual exclusion problem. *Infor-
            mation Processing Letters,* 12(3):115–116, June 1981.

[PM96]   W. Peng and K. Makki. Petri nets and self-stabilization of communi-
            cation protocols. *Informatica,* 20:113–123, 1996.

[Ray88]   M. Raynal. *Distributed Algorithms and Protocols.* Wiley Series in
            parallel computing. J. Wiley and Sons, 1988.

[Ray89]   K. Raymond. A tree-based algorithm for distributed mutual exclusion.
            ACM Transactions on Computer Systems, 7(1):61–77, February 1989.

[Rei85]   W. Reisig. *Petri Nets,* volume 4 of *EATCS Monographs on Theoretical
            Computer Science.* Springer-Verlag, Berlin, 1985.

[Rei95]   W. Reisig. Petri net models of distributed algorithms. In Jan van
            Leeuven, editor, *Computer Science Today. Recent Trends and Devel-
            opments,* volume 1000 of *LNCS Lecture Notes in Computer Science,*
            pages 441–454. Springer-Verlag, Berlin, 1995.

[Rei96a]   W. Reisig. Interleaved progress, concurrent progress, and local
            progress. In D. A. Peled, V. R. Pratt, and G. J. Holzmann, edi-
            tors, *Partial Order Methods in Verification,* volume 29, pages 24–26.
            DIMACS Series in Discrete Mathematics and Theoretical Computer
            Science, American Mathematical Society, 1996.

[Rei96b]   W. Reisig. Modeling and verification of distributed algorithms. In
            U. Montanari and V. Sassone, editors, *CONCUR 96: Concurrency
            Theory,* volume 1119 of *LNCS Lecture Notes in Computer Science,*
            pages 79–95. Springer-Verlag, 1996.

[RH90]   M. Raynal and J.-M. Helary. *Synchronization and Control of Distrib-
            uted Systems and Programs.* Wiley Series in parallel computing. J.
            Wiley and Sons, 1990.

[RK97]   W. Reisig and E. Kindler. Verification of distributed algorithms with
            algebraic Petri nets. In C. Freksa, M. Jantzen, and R. Valk, editors,
            Foundations of Computer Science — Potential, Theory, Cognition,
            volume 1337 of *LNCS Lecture Notes in Computer Science,* pages 261–
            270. Springer-Verlag, 1997.

[Roz86]   G. Rozenberg. Behaviour of elementary net systems. In W. Brauer,
            W. Reisig, and G. Rozenberg, editors, *Petri Nets: Central Models and
            Their Properties,* volume 254 of *LNCS Lecture Notes in Computer
            Science,* pages 60–94. Springer-Verlag, 1986.

[Sch97]   F. B. Schneider. *On Concurrent Programming.* Springer, 1997.

[Seg83]   A. Segall. Distributed network protocols. *IEEE Transactions on In-
            formation Theory,* IT 29-1:23–35, 1983.

[SF86]   N. Shavit and N. Francez. A new approach to detection of locally
            indicative stability. In L. Kott, editor, *Proceedings of the 13th ICALP*,
            volume 226 of *LNCS Lecture Notes in Computer Science*, pages 344–
            358. Springer-Verlag, 1986.

[Tel91]   G. Tel. *Topics in Distributed Algorithms*, volume 1 of *Cambridge
            International Series on Parallel Computation*. Cambridge University
            Press, Cambridge, U.K., 1991.

[Tel94]   G. Tel. *Introduction to Distributed Algorithms*. Cambridge University
            Press, Cambridge, U.K., 1994.

[Völ97]   H. Völzer. Verifying fault tolerance of distributed algorithms for-
            mally: A case study. Informatik-Berichte 84, Humboldt-Universität
            zu Berlin, May 1997. (appears in: Proceedings of CSD98, Interna-
            tional Conference on Application of Concurrency to System Design,
            Aizu-Wakamatsu City, March 1998, IEEE Computer Society Press).

[Val86]   R. Valk. Infinite behaviour and fairness. In W. Brauer, W. Reisig, and
            G. Rozenberg, editors, *Petri Nets: Central Models and Their Proper-
            ties*, volume 254 of *LNCS Lecture Notes in Computer Science*, pages
            377–396. Springer-Verlag, 1986.

[Wal95]   R. Walter. *Petrinetzmodelle verteilter Algorithmen*. PhD thesis,
            Humboldt-Universität zu Berlin, Institut für Informatik. Edition Ver-
            sal, vol. 2. Bertz Verlag Berlin, 1995.

[Wal97]   R. Walter. The asynchronous stack revisited: Rounds set the twilight
            reeling. In C. Freksa, M. Jantzen, and R. Valk, editors, *Foundations of
            Computer Science*, volume 1337 of *LNCS Lecture Notes in Computer
            Science*, pages 307–312. Springer-Verlag, 1997.

[WWV+98]   M. Weber, R. Walter, H. Völzer, T. Vesper, W. Reisig, S. Peuker,
            E. Kindler, J. Freiheit, and J. Desel. DAWN: Petrinetzmodelle zur
            Verifikation verteilter Algorithmen. Informatik-Bericht 88, Humboldt-
            Universität zu Berlin, 1998.

Springer
and the
environment

At Springer we firmly believe that an international science publisher has a special obligation to the environment, and our corporate policies consistently reflect this conviction.

We also expect our business partners – paper mills, printers, packaging manufacturers, etc. – to commit themselves to using materials and production processes that do not harm the environment. The paper in this book is made from low- or no-chlorine pulp and is acid free, in conformance with international standards for paper permanency.

 Springer